T0282418

Linked by Blood:
Hemophilia and AIDS

Frontispiece: Ruth Seeler, MD (back row, third from left) and boys attending hemophilia summer camp, photographed prior to the AIDS epidemic. Many of these campers subsequently became infected by the human immunodeficiency virus (HIV) and succumbed to AIDS. *Courtesy of Bob Robinson, Executive Director, Bleeding Disorders Alliance Illinois.*

Linked by Blood:
Hemophilia and AIDS

David Green, MD, PhD
Professor of Medicine Emeritus,
Feinberg School of Medicine of Northwestern University,
Chicago, IL, United States

AMSTERDAM • BOSTON • HEIDELBERG • LONDON
NEW YORK • OXFORD • PARIS • SAN DIEGO
SAN FRANCISCO • SINGAPORE • SYDNEY • TOKYO

Academic Press is an imprint of Elsevier

Academic Press is an imprint of Elsevier
125 London Wall, London EC2Y 5AS, UK
525 B Street, Suite 1800, San Diego, CA 92101-4495, USA
50 Hampshire Street, 5th Floor, Cambridge, MA 02139, USA
The Boulevard, Langford Lane, Kidlington, Oxford OX5 1GB, UK

British Library Cataloguing-in-Publication Data
A catalogue record for this book is available from the British Library.

Library of Congress Cataloging-in-Publication Data
A catalog record for this book is available from the Library of Congress.

ISBN: 978-0-12-805302-7

For Information on all Academic Press publications
visit our website at https://www.elsevier.com/

Working together
to grow libraries in
developing countries

www.elsevier.com • www.bookaid.org

Publisher: Mica Haley
Acquisition Editor: Tari K. Broderick
Editorial Project Manager: Jeffrey L. Rossetti
Production Project Manager: Chris Wortley
Designer: Mark Rogers

Typeset by MPS Limited, Chennai, India

Dedication

This book is dedicated to my patients and their families, whose courage and fortitude inspired me to undertake this work.

Contents

Preface

While I was in college, I spent two summers working in a laboratory devoted to the study of blood coagulation. It was fascinating to observe how liquid blood underwent clotting within a few minutes of removal from a vein. But when blood was taken from patients with hemophilia, it remained liquid for an hour or more. The laboratory investigated the factors that promoted or hindered coagulation in normal and hemophilic blood, and I continued part-time participation in this laboratory research during medical school, residency, and fellowship training.

During my fellowship year, I was assigned to care for one of our oldest and most severely hemophilic patients. This 49-year-old man had a history of lifelong bleeding episodes, beginning with a major hemorrhage after a tonsillectomy at age 10. Over the years, he had repeated joint hemorrhages that left him crippled; he had been bedridden for the 7 years prior to his current admission. Two cousins and a nephew died of hemophilic bleeding, and his two brothers and another nephew had hemophilia. He was brought to our hospital on this occasion because blood was seen in his stool. Because bleeding was quite brisk, he was given several units of blood and plasma; although this slowed the bleeding, he still required 2 pints of blood per day to keep up with the continuing losses. This situation persisted for the next 2 months, and during this time his body became quite swollen. There was edema fluid under his skin, in his lungs and abdomen, and even in the walls of his intestines. Laboratory tests showed a pattern consistent with severe malnutrition; the protein level in his blood was extremely low. At this point, our team wondered if all the edema fluid might be contributing to the bleeding by restricting the healing of his bowel, and we decided to focus our attention on correction of the protein deficiency. We increased the dietary protein and gave additional protein intravenously. After 2 weeks, his blood protein level was much improved and the edema subsided. He passed the first normal stool since his admission more than 2 months previously, his blood count increased, and transfusions were stopped. Soon thereafter he was discharged from the hospital; at follow-up 3 months later, there had been no recurrence of the bowel bleeding.[1]

---

1. This case history was reported as: Green D, Geisler PH, Erslev AJ. Protein-losing enteropathy complicating prolonged bleeding in hemophilia. Ann Intern Med 1964;61:928–32.

Although our patient's response to the protein therapy was gratifying, why wasn't the problem of malnutrition addressed sooner? The continuing focus on hemophilia and inattention to the patient's general condition probably prolonged his hospitalization; bleeding stopped only after the nutritional deficiency was corrected. This error might have been avoided by recognition that more than one illness might be in play in a given individual. This concept of multiple concurrent illnesses typified the situation in the 1980s, when hemophiliacs were stricken with AIDS and infections could appear in almost any organ. Treating the entire patient, and not just one or another organ system, was critical for patient survival and is recognized as consistent with the best medical practice.

The satisfactory outcome achieved by this patient seen during my fellowship year strengthened my resolve to become a hematologist and to provide care for persons with bleeding and clotting disorders. After my fellowship, I served 2 years in the US Navy and then received a Public Health Service Special Fellowship for advanced training at the Blood Coagulation Unit in Oxford, England. This Unit was the epicenter of clotting research and hemophilia treatment in the 1960s. Equipped with the knowledge and experience gained during this second fellowship year, I moved back to the United States and assumed the management of persons with hemophilia at a large urban medical center.

In the early 1970s, most physicians were not familiar with the intricacies of blood coagulation and were not eager to undertake the care of patients with a relatively rare but serious disorder. When a bleeding hemophiliac arrived at their local hospital emergency department, they usually asked us to accept the patient in transfer. Patients soon learned to bypass these emergency departments and come to us directly. Our hemophilia service rapidly expanded; within 1 year, more than 100 patients were registered for diagnostic testing and treatment.

As a physician in an academic medical practice, I regularly read a variety of medical journals and pamphlets and attended meetings and conferences. I saved articles and publications related to hemophilia and other coagulation disorders. When the first cases of AIDS in hemophiliacs were reported, I filed the published works in a folder labeled Hemophilia & AIDS that eventually grew to contain almost 300 journal articles, letters, and other communications. It included several items from Jeanne Lusher, MD, our Regional Representative of the Medical and Scientific Advisory Council (MASAC) of the National Hemophilia Foundation (NHF). In 1982, Dr Lusher began to distribute the MASAC meeting minutes and NHF communications. The latter were in the form of chapter advisories and medical bulletins, initially called *Hemophilia NewsNotes* and subsequently called the *Hemophilia Information Exchange*. In addition, she sent copies of the Centers for Disease Control (CDC) *Morbidity and Mortality Weekly Reports*, Food and Drug Administration bulletins, Council of Community Blood Centers newsletters, and other items about hemophilia and AIDS. Another important source of information was *The Hemophilia Bulletin*, a newsletter published by Carol K. Kasper, MD, a hemophilia physician at the Los Angeles Orthopaedic Hospital. All of these publications, as well as newspaper articles, books, Internet sites, letters from colleagues, and my own personal experience, are the sources for the material presented in this volume.

Blood clotting factor concentrates were used for the prevention and control of bleeding. They were very effective and could be used at home as well as in the hospital. When given prior to surgery, these concentrates made it possible to perform surgical correction of deformities and other operative procedures in hemophliacs, and they enabled previously crippled individuals to resume schooling and achieve gainful employment. People with bleeding disorders were able to lead relatively normal lives, and seeing them improve was very gratifying to their physicians. This was the state of affairs in the early 1980s, prior to the AIDS epidemic.

David Green, MD, PhD
Hematology/Oncology
Suite 1020
645 N Michigan Avenue
Chicago, IL 60611
United States
September 2015

Acknowledgments

I am extremely grateful to my wife, Theodora, whose advice, suggestions, and unflagging encouragement were major contributors to this book. Also, our daughter, Jo Ellen G. Kaiser, provided valuable and constructive criticism, and Jennifer Green (no relation) critiqued some of the chapters and gave very helpful advice. I thank Sandy Harris, our Bleeding Disorders Coordinator, for refreshing my memory about patients and events, Frank Palella, MD, for reviewing the chapter "The Human Immunodeficiency Virus" for accuracy and completeness, and Glenn Ramsey, MD, for providing information about donor recruitment and blood donation. I also acknowledge my colleagues Joanne Goldsmith, MD, John Phair, MD, and Nasim Rana, MD, who provided outstanding care for our patients with hemophilia and AIDS during the tumultuous decade of the 1980s.

Introduction

Hemophilia, an inherited disorder of blood coagulation, literally means "love of blood." It is an appropriate name for this condition because people with hemophilia are dependent on the clotting proteins present in blood to prevent bleeding. After World War II, transfusion of blood from healthy donors was found to be effective for controlling hemophilic hemorrhages. Subsequently, refinements in the processing and packaging of blood clotting factors by pharmaceutical companies during the 1950s and 1960s led to the widespread marketing of clotting factor concentrates. These blood derivatives improved the quality of life and lengthened the life span of hemophiliacs.

By the 1970s, most persons with hemophilia had become completely dependent on commercial clotting factor concentrates; however, in the early 1980s, a mysterious illness began to sicken and kill hemophilic men and boys, as well as other recipients of blood transfusions. This sickness first appeared a few years earlier in homosexual men and drug addicts in San Francisco and New York and was called gay-related immune deficiency (GRID) [1]. By May 1982, the Centers for Disease Control and Prevention (CDC) reported 355 GRID cases and 136 deaths, mainly in persons residing in New York City (45%) and California (20%). The disease was characterized by an increased susceptibility to infection and became known as acquired immunodeficiency syndrome, or AIDS, and soon reached epidemic proportions. By December 1983, there were 3000 cases and 1283 deaths [2]. Even though large numbers of gay men were being stricken, the public response was muted; there was no major campaign to eradicate the disease, probably because of long-standing antagonism toward persons with alternative sexual orientations. These individuals were often called "sexual deviants" and, as late as the 1960s, the discovery that an employee was gay resulted in his or her firing [3]. Many thought that AIDS was retribution for the uninhibited sexual lifestyle attributed to gays. The lack of a strong response to the epidemic had important ramifications: political advocacy and financial support for combating the disease were limited and the medical establishment was given few tools to fight the disease.

Media attention and research funding gradually increased as more hemophiliacs and other heterosexual persons developed AIDS and died. As Randy Shilts noted, "The gay plague got covered only because it finally had struck people who counted, people who were not homosexuals" [4]. But the increased media coverage had the perverse effect of provoking widespread fear of contagion and avoidance of those suspected of being infected. Soon, hemophiliacs and their families were subject to the same hostile attitudes that generally characterized the public response to homosexuality; they were banned from schools, discriminated against in the workplace, and became social outcasts. Eventually, medical researchers discovered that the disease was due to a virus and methods were developed to contain the infection. As the number of new cases diminished, people became less fearful and were more accepting of persons with hemophilia. Late in 1983, the virus that caused AIDS was identified, and this led to the development of methods to diagnose the disease and prevent the spread of this virus.

The AIDS epidemic was a tragic event of epic proportions, initially affecting a few select groups but then spreading to the general population. Recounting the history of the epidemic is valuable because it highlights the responses of individuals, organizations, communities, and eventually government to the threat of contagion. It required mobilization of financial resources, key scientific discoveries, and major efforts by blood banks and pharmaceutical companies to control the infection. These activities required leaders to overcome established prejudicial attitudes and mobilize public opinion to support education and research. Although there has been significant progress in the 30-plus years since the epidemic first appeared in the United States, HIV/AIDS is still a major cause of disease and death worldwide, and much work remains to be done.

This book begins with a look at the impact that the disease had on hemophiliacs and their families. It then describes the actors in this drama, hemophilia, blood, and HIV, and relates how blood that was inadvertently contaminated by HIV was used to prepare the clotting factor concentrates infused by hemophiliacs. In ensuing chapters, the eventual dimensions of the epidemic, the measures for infection control, and the interventions to improve the safety of blood products are reviewed. In addition, the ways that doctors and patients acted in response to the mounting numbers of infected persons are recounted, relying on published accounts and the author's personal experience.

After the HIV epidemic had run its course, a Committee was appointed by the US Institute of Medicine to examine factors that might have delayed implementation of steps to limit the transmission of the virus by blood and blood products; the outcome of that investigation is summarized and analyzed. Subsequent chapters examine the factors that enabled the spread of the virus and the contributions by organizations and individuals that mitigated the epidemic. The penultimate chapter describes how the care of people with hemophilia has evolved over the ensuing three decades, and the final chapter presents suggestions for measures to improve blood safety, availability, and access to essential drugs and blood products.

Revisiting AIDS and hemophilia provides important lessons for managing current and future encounters with virulent organisms. Recent contagions causing worldwide distress are Ebola viral disease (EVD), SARS, MERS, and Zika viruses. EVD was ignored for many months and only came to prominence when it was recognized that it could spread from West Africa to every continent. Similarly, AIDS killed thousands of homosexual men before the infection of a movie star, Rock Hudson, awakened the public to its dangers. A massive informational campaign was required to contain AIDS, and a similar effort will be necessary to limit the transmission of other lethal viruses. Education about the cause of AIDS persuaded infected individuals to reject conspiracy theories and accept treatment; similarly, efforts to enlighten those affected by Ebola and other viruses will encourage them to seek medical assistance. Educating the public helped mitigate the stigma attached to AIDS, and it might curb the irrational fears that have led some to avoid anyone or anything even remotely connected with EVD, even though this infection is not spread by indirect contact.

A final lesson to be learned from these epidemics is that the pharmaceutical industry rarely expends resources on products that are not deemed profitworthy; for example, research to develop safer products for hemophiliacs languished prior to the AIDS epidemic, and a vaccine to prevent EVD sat on the shelf for years [5]. Governments must demand that manufacturers devote a greater percentage of their research and development (R&D) to the formulation of drugs and vaccines for diseases that affect the economically disadvantaged as well as the wealthy. Furthermore, products that are approved by the Food and Drug Administration (FDA) must be accessible to all who need them; this will require government controls on the prices of treatments essential for the health of the population. The experience with AIDS and hemophilia has shown that education, research, and enlightened regulatory activity can restore the health and productivity of ravaged populations.

REFERENCES

[1] Shilts R. And the band played on. New York: St Martin's Press; 1987, p. 152.
[2] Shilts R. And the band played on. New York: St Martin's Press; 1987, p. 401.
[3] Apuzzo M. Uncovered papers show past government efforts to drive gays from jobs. NY Times May 21, 2014:A14.
[4] Shilts R. And the band played on. New York: St Martin's Press; 1987, p. 126.
[5] Grady D. Ebola vaccine, ready for test, sat on the shelf. NY Times October 24, 2014.

Chapter 1

The Impact of AIDS on Hemophilia

During the 1970s, most individuals with hemophilia began to use clotting factor concentrates because these products dramatically improved their lives and enabled them to enter the mainstream of society. They and their physicians considered these therapeutic materials to be indispensable for the modern treatment of hemophilia. However, the preparation of concentrate required the use of plasma from thousands of donors, and the inclusion of a single plasma donation from one person infected by a virus could contaminate an entire lot of concentrate.

The 1970s also saw the emergence of the gay liberation movement and the appearance of bathhouses and sex clubs that allowed homosexuals to engage in unlimited sexual activity. A single individual infected by human immunodeficiency virus (HIV) could spread this virus to hundreds of partners. Homosexuals were a major source of the plasma used to prepare clotting factor concentrates; during the 1970s, they donated 5–9% of the blood collected by the Irwin Memorial Blood Bank in San Francisco [1]. Concentrate potentially contaminated by HIV was widely used for the treatment of hemophiliacs in the United States and shipped internationally.

Between 1980 and 1992, the Centers for Disease Control (CDC) reported that at least 1928 of the nation's estimated 15,000 hemophiliacs were infected by HIV, and 1654, or 1 in 10, died of acquired immunodeficiency syndrome (AIDS) [2]. In Canada, 660 of 2427 (27.2%) hemophiliacs became HIV-positive, and 406 died [3]. The worldwide decimation of hemophilia communities was a consequence of the convergence of concentrate therapy for hemophiliacs and HIV contamination of the blood supply that occurred in the United States in the late 1970s.

The impact of AIDS on hemophiliacs and their families was enormous, affecting not only health but also social relationships, education, and work. Prior to the epidemic, people with this disorder lived in the "golden era" of hemophilia [4]. Medical advances had resulted in the availability of commercial clotting factor concentrates that could be infused at home by the person with hemophilia or a family member. These blood products could arrest serious

Linked by Blood: Hemophilia and AIDS. DOI: http://dx.doi.org/10.1016/B978-0-12-805302-7.00001-X

bleeding within minutes, alleviating pain and preventing disability. For the first time, hemophiliacs could maintain satisfactory school attendance, select rewarding careers, and become productive members of society.

AIDS changed everything. People who were unaware they were infected by HIV donated blood, and this HIV-contaminated blood was used to manufacture the clotting factor concentrates used by hemophiliacs. Hundreds of hemophilic men and boys became infected by the virus, and although they had learned to cope with painful muscle and joint hemorrhages before the AIDS epidemic, they now encountered a disease that could affect any organ in their bodies. Almost all developed fatigue, weight loss, and swollen lymph glands. Pneumonia was common, as were fevers and diarrhea. Clinic visits and hospitalizations became frequent, causing major disruptions in work, school, and family life.

The experience of hemophiliacs receiving their care at the Northwestern Center for Bleeding Disorders was fairly typical. In November 1983, 43 hemophiliacs were enrolled there; 33 (77%) had symptoms and laboratory abnormalities consistent with HIV infection [5]. To accommodate these infected individuals, a combined hemophilia and HIV clinic was established to deal with the myriad of new clinical disorders associated with AIDS. A "one-stop" service was provided; each patient was interviewed, examined, and treated by a hematologist and an infectious disease specialist. A nurse-coordinator and social worker rounded out the team; this enabled each patient to receive comprehensive care for hemophilia and HIV infection.

The destruction of the immune system that accompanied HIV infection predisposed to several other disorders, such as decreased blood platelets, cancer of the lymph glands, and progression of chronic viral hepatitis. A low platelet count was observed frequently and was particularly harmful because platelets are needed to prevent bleeding, and their loss can aggravate hemophilic bleeding. Furthermore, the customary treatment for low platelets, a steroid drug, increases the risk of infection, especially in highly susceptible HIV-infected persons. Therefore, the treatment of these patients was problematic. Other therapies, such as intravenous gamma globulin, were effective but were an option for only a few patients because of their high cost [6].

Cancer of the lymph glands occurred in 5.5% of HIV-positive hemophiliacs and was eventually fatal in many patients despite an initial response to chemotherapy [7]. Occasionally, previously rarely seen disorders were encountered. For example, a young man with hemophilia and HIV infection developed an atypical form of tuberculosis. Multiple courses of powerful antibiotics controlled the tuberculosis, but he began experiencing severe fluid retention, with extreme swelling of his face, abdomen, arms, and legs. He became bedridden and eventually died of pneumonia. At autopsy, heavy deposits of an abnormal fibrous protein called amyloid were found in his kidneys. Amyloid levels are increased in the blood of AIDS patients (and amyloid disease has been reported in monkeys with AIDS). Amyloid deposits in the kidneys damage the normal filtering apparatus (the glomeruli and tubules shown in gold in Fig. 1.1),

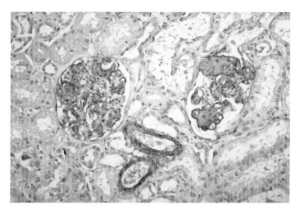

FIGURE 1.1   Amyloid (gold) deposited in the glomeruli and tubules of the patient's kidney. *Photograph courtesy of the author.*

permitting the loss of normal proteins from the blood and causing fluid retention; this was the cause of the patient's swelling [8].

Almost all of the patients attending the Northwestern Hemophilia Center were exposed to the hepatitis viruses that contaminated the clotting factor concentrates available at the time. Laboratory tests for these individuals revealed low-grade inflammation of the liver. However, coinfection with HIV resulted in rapid progression of the inflammation, often culminating in death from liver failure. Multicenter studies report that HIV infection increases the risk of end-stage liver disease more than eightfold [9,10]. Some of these patients were candidates for liver transplantation, but even in those with a successful transplant, long-term treatment with powerful antirejection medications was required. These drugs were expensive and had a variety of unpleasant side effects.

Other patients developed infections with organisms that were previously considered exotic. An HIV-infected patient had acute onset of severe headaches and visual impairment, followed by muscle paralysis. A brain biopsy showed progressive multifocal leukoencephalopathy (PML), a rare infectious disease of the brain caused by the J-C virus. There was no treatment for this infection, and the patient died within a few weeks. Several years later, his 26-year-old hemophilic nephew developed AIDS and also died of PML.

Hemophilic men were warned that sexual exposure to HIV posed a risk to their wives. Accepting a prohibition on unprotected sex was especially difficult for young couples wanting to start families. In vitro fertilization was not an option; at the time, no method for sterilizing semen was available that would kill the virus but spare the sperm.[1] This meant that couples had to delay starting families indefinitely, which was a great disappointment to many.

---

1. Sperm washing is now an effective technique for ensuring a safe pregnancy when the male donor is HIV-positive (Savasi V, et al. Human Reprod 2007;22:772–7).

Even the treatment of HIV was problematic for some patients. Antiviral drugs often caused nausea, fatigue, and headache, as well as anemia and discoloration of the skin and nails. Another distressing side effect was the accumulation of large amounts of fatty tissue around the neck and upper back. This condition, called lipodystrophy, could be very disfiguring. Although most patients were able to continue working despite the ugly burden, some needed surgery because the mass of tissue became so large that they could no longer button their collars.

SOCIAL IMPACT

When AIDS first came to public attention, it was considered to be a medical oddity because the disease outbreak was largely restricted to the male homosexual communities of New York, Los Angeles, and San Francisco [11]. However, as the number of cases rapidly escalated, AIDS patients soon filled hospital wards in these cities at an alarming rate. Public health workers were concerned that the epidemic would spread from homosexuals to other population groups, and that those who became infected would be subject to social discrimination. Homosexuals have often been targets of intolerance; when AIDS, a disorder predominantly affecting homosexuals, was discovered in hemophiliacs, they encountered similar prejudicial attitudes.

Because the risks of contact with HIV-infected persons were unknown at the time, the National Institutes of Health (NIH) issued broad warnings about exposure to saliva, tears, and other bodily fluids. As an unintended consequence of this action, the general population came to fear proximity to persons with hemophilia and sought to avoid direct and indirect contact with them and their family members. Ryan White, a young man with hemophilia and AIDS, reported that people would not sit near him in restaurants and refused to shake his hand, even in church. He was required to seek separate drinking fountains and restrooms and use disposable eating utensils and trays; in fact, some restaurants even threw away his dishes [12]. After White was diagnosed with AIDS, many parents and teachers in his community feared that he might infect other children, and they signed a petition encouraging school leaders to ban him from returning to his classroom. After an 8-month legal battle, he was finally allowed to attend school—only to be shunned by his classmates and harangued with epithets such as, "we know you're queer." White's supporters were called homosexuals and threatened with death. Windows in his home were broken, the tires of the family car were slashed, and he and his family had to relocate to another town [13]. After these events were reported in the press, Ryan White was interviewed frequently and became a spokesperson for hemophilia and AIDS education. He appeared on national television with celebrities such as Elton John and was quoted in newspapers. He died of AIDS in 1990 at age 19.

In Florida, the local school board refused to allow the three Ray brothers, all of whom had hemophilia and HIV, to attend their public school. The family mounted a legal appeal, but 1 week after the school board action was overturned by a federal court, the Ray family home was burned to the ground and they were forced to leave the community. The two older boys died of AIDS, one at age 15 and the other at age 22, and their father made a failed attempt at suicide soon after their deaths [14].

Adults with hemophilia and AIDS were also subject to discrimination, mainly in the workplace. Even those free of HIV infection had limited vocational choices; for example, they could not engage in construction work, jobs requiring prolonged standing and lifting, or activities that might provoke bleeding. People with AIDS were barred from working as food handlers, beauticians, and other jobs involving close personal contact, but employment discrimination extended even further. Many employers were reluctant to hire HIV-infected persons because they feared that other workers would refuse to labor alongside them. As a consequence, men and boys with hemophilia hid their HIV status from their friends, coworkers, and associates. Nevertheless, they and their families often became social outcasts and suffered severe economic and mental distress. Antagonism toward HIV-infected persons persisted for many years, even after it was unequivocally demonstrated that the virus was not acquired by casual contact.

The impact of AIDS on family life was revealed by a 1985 survey of 935 hemophiliacs in the Netherlands; the survey found that 65% were preoccupied with AIDS and 31% had feelings of depression [15]. Patients often abandoned their usual therapeutic regimens because of fears that the clotting factor concentrates might be contaminated with HIV. Patients with a hemorrhage delayed infusing clotting factor or used smaller, less effective amounts in the hope that bleeding would spontaneously subside; even those with frequent episodes of bleeding stopped taking preventive doses of concentrate. These actions resulted in more severe hemorrhages, exacerbated pain and disability, and led to deterioration in patients' general health. Some men were in a state of denial about the possibility of being infected by the virus and failed to use condoms, placing their wives at risk. They also put themselves at risk by neglecting to use gloves or special containers for handling needles and syringes. In summary, the hemophilia community was under tremendous stress, and almost every aspect of family life was affected. Children were barred from school attendance and shunned by classmates, men with hemophilia became unemployed, conjugal relations were abridged, and many hemophiliacs were abandoned by friends and associates.

MEDICAL IMPACT

Individuals with hemophilia had to contend with the treatment of bleeding as well as HIV and hepatitis. During the 1980s, the Northwestern clinic eventually served 92 patients with hemophilia; 72 were deficient in clotting factor VIII and 20 were deficient in clotting factor IX (the two common types of hemophilia; see Chapter 2). Testing for HIV exposure showed that 54% of the factor VIII patients and 55% of the factor IX patients were HIV-positive [16]. It was suspected that these patients had been infected by the blood products they used to treat hemorrhages. When safer blood products became available in 1985, a letter was sent to all patients advising that they should only use the new heat-treated concentrates, carefully dispose of items that might be contaminated with blood, and maintain regular visits to the clinic (Fig. 1.2).

NORTHWESTERN UNIVERSITY
CHICAGO, ILLINOIS 60611

DAVID GREEN, M.D., Ph.D.
PROFESSOR OF MEDICINE
DIRECTOR, ATHEROSCLEROSIS
CHIEF OF MEDICINE
REHABILITATION INSTITUTE OF CHICAGO

June 24, 1985

Dear Patient:

To prevent the spread of AIDS infection, users of concentrate should

1. Use only heat-treated concentrates.

2. Dispose of infusion supplies in solid-walled containers until the contents can be incinerated.

3. Blood spilled during infusions should be cleaned with diluted household bleach.

4. Do not share implements such as toothbrushes or razors with anyone else.

5. Use condoms for sexual intercourse.

6. Finally, do not delay coming in for your annual visits, and do not hesitate to phone if you have problems or questions.

Sincerely,

David Green, M.D., Ph.D.

DG/el

THE McGAW MEDICAL CENTER OF NORTHWESTERN UNIVERSITY

FIGURE 1.2   Letter to patients.

HIV infection dramatically increased the death rate among hemophiliacs. Sixteen percent of the Clinic's patients died during the next 20 years: six from infections, five from cancer, two after liver transplantation, and two from bleeding associated with injuries. The median age of patients at the time of death was only 49 (range, 26–69). The Multicenter Hemophilia Cohort Study recorded the causes of death for 1028 participants between 1982 and 1995 [17]. The death rate was 13-fold higher in hemophiliacs infected by HIV (36.8% vs 2.8%). A study of hemophilia from the Netherlands concluded that the life expectancy of patients with hemophilia would almost equal that of the general population if deaths from HIV and hepatitis were excluded [18].

ECONOMIC IMPACT

The financial burden of hemophiliacs increased exponentially during the decade of the HIV epidemic (1982–92). The cost per unit of concentrate increased from $0.06–0.09 in 1982 to $0.47–0.60 in 1987 and $0.70–0.90 in 1992. Between 1982 and 1987, pasteurization and other viral inactivation procedures were implemented; in 1992, recombinant products became available [19].

In 1983, the average patient with severe hemophilia was using 30,000–50,000 units per year [20], and costs increased from $1800–4500 in 1982 to $16,500–30,000 in 1987, and to $21,000–50,000 in 1992. Today, the annual cost of blood products for the treatment of bleeding episodes is often more than $100,000; it can be $300,000 to prevent bleeding, which requires doses two or three times per week.

In addition to greatly increased expense for clotting factor, hemophiliacs had to visit healthcare facilities more often to receive HIV screening, counseling, and other services. Those infected with AIDS had the additional expense of hospitalizations for infections, tests to monitor immune status, and costly anti-HIV medications. In 1998, the Ricky Ray Hemophilia Relief Fund Act authorized $750 million to support one-time payments of $100,000 to HIV-infected persons with hemophilia treated with clotting concentrates between 1982 and 1987 [21]. Their HIV-infected spouses were also eligible to receive payment. In 1990, Congress authorized the Ryan White HIV/AIDS Program to fill gaps in insurance coverage for HIV-related services for more than half a million infected people each year. The Act also supported AIDS education and training centers, dental programs, and minority AIDS initiatives. At the state level, new programs provided additional need-based support for hemophilia patients. These measures helped soften the financial burden of hemophilia/HIV care for many individuals.

Perhaps the most important legislation affecting persons with hemophilia is the Affordable Care Act (ACA), enacted in 2010 [22]. It includes a prohibition on denying coverage to children younger than age 19 years based on preexisting conditions and provides new coverage options to individuals who have been uninsured for at least 6 months because of preexisting conditions. In the past, many families were unable to obtain coverage for their hemophilic children;

when these children became adults, they were unable to obtain insurance. The Act also eliminates lifetime limits on insurance coverage for essential benefits (such as clotting factor concentrates) and bans annual dollar limits on hospital stays. This is a significant benefit because the lifetime healthcare expenses of persons with hemophilia often exceed the $1 million limits commonly set by insurance companies. In addition, the ACA establishes an external review process for appealing insurance company claims decisions.

The impact of the AIDS epidemic on the hemophilia community led to considerable litigation. The most prominent targets were the manufacturers of clotting factor concentrates, who were accused of product safety failures. These lawsuits were usually not successful due to the general lack of knowledge about HIV at the time people became infected. For example, this defense was used in a case against Georgetown and the American Red Cross involving transfusion-transmitted AIDS; the Court granted motions for summary judgment and dismissed all complaints [23]. Other legal actions against commercial manufacturers of factor concentrates were settled by payments of $100,000 to each HIV-infected person with hemophilia and any other person they had infected, matching the compensation provided by the Ricky Ray Act. In other countries, both civil and criminal litigation led to monetary awards to hemophiliacs and imprisonment of health officials and physicians who had knowingly dispensed HIV-contaminated clotting factor concentrates when safer products were available [24].

In summary, HIV infection impacted every aspect of the lives of people with hemophilia. By 1985, intensive efforts led to improvement in the safety of the blood supply and cessation of new infections in people with hemophilia. Educational initiatives by the CDC and the World Hemophilia AIDS Center informed the hemophilia population about measures required to decrease the sexual transmission of HIV, and spousal infection became very infrequent [25]. However, for those hemophiliacs infected between 1981 and 1985, local discrimination and stigmatization remained common occurrences. Writing in 1987, Margaret Hilgartner, a physician responsible for the care of children with hemophilia, called for society to find a solution to the social problems created by AIDS [26]. She stressed the need for an intensive campaign to educate every member of the community about the transmission and natural history of the disease.

KEY POINTS

- HIV infection had serious adverse effects on the social, medical, and financial health of hemophiliacs.
- National Institutes of Health warnings about exposure to saliva, tears, and other bodily fluids had the unintended consequence of people fearing proximity to hemophiliacs, and this resulted in widespread social isolation.
- Office visits and hospitalizations by hemophiliacs became more frequent, concomitant infections were common and severe, and life expectancy was dramatically decreased.

- The financial liability of hemophiliacs increased exponentially between 1982 and 1992, although government legislative actions provided some relief.
- Litigation was initiated against blood bankers, pharmaceutical firms, and government health officials in Europe and North America, who were accused of knowingly dispensing contaminated blood and blood products.

REFERENCES

[1] Shilts R. And the band played on. New York, NY: St Martin's Press; 1987, p. 199.

[2] Resnik S. Blood saga. Berkeley, CA: University of California Press; 1999, p. 242.

[3] Arnold DM, Julian JA, Walker IR. Mortality rates and causes of death among all HIV-positive individuals with hemophilia in Canada over 21 years of follow-up. Blood 2006;108:460–4.

[4] Resnik S. Blood saga. Berkeley, CA: University of California Press; 1999, p. 292.

[5] Goldsmith JM. Immunologic abnormalities in patients with hemophilia: association with use of factor concentrates. Arch Intern Med 1985;145:431–4.

[6] Pollak AN, Janinis J, Green D. Successful intravenous immune globulin therapy for human immunodeficiency virus-associated thrombocytopenia. Arch Intern Med 1988;148:695–7.

[7] Ragni MV, Belle SH, Jaffe RA, et al. Acquired immunodeficiency syndrome-associated non-Hodgkin's lymphomas and other malignancies in patients with hemophilia. Blood 1993;81:1889–97.

[8] Cozzi PJ, Abu-Jawdeh GM, Green RM, Green D. Amyloidosis in association with human immunodeficiency virus infection. Clin Infect Dis 1992;14:189–91.

[9] Goedert JJ, Eyster ME, Lederman MM, et al. End-stage liver disease in persons with hemophilia and transfusion-associated infections. Blood 2002;100:1584–9.

[10] Posthouwer D, Makris M, Yee TT, et al. Progression to end-stage liver disease in patients with inherited bleeding disorders and hepatitis C: an international, multicenter cohort study. Blood 2007;109:3667–71.

[11] Groopman JE, Detsky AS. Epidemic of the acquired immunodeficiency syndrome: a need for economic and social planning. Ann Intern Med 1983;99:259–61.

[12] White R. Testimony before the President's Commission on AIDS; 1988.

[13] Johnson D. Ryan White dies of AIDS at 18; his struggle helped pierce myths. NY Times 1990;April 9.

[14] Ray brothers. Accessed from: Wikipedia; May 15, 2014.

[15] Rosendaal FR, Smit C, Varekamp I, Brocker-Vriends A, Suurmeijer TPBM, Briet E. AIDS and haemophilia. Haemostasis 1988;18:73–82.

[16] Goldsmith JM, Variakojis D, Phair JP, Green D. The spectrum of human immunodeficiency virus infection in patients with factor IX deficiency (Christmas disease). Am J Hematol 1987;25:203–10.

[17] Goedert JJ. Mortality and haemophilia. Lancet 1995;346:1425–6.

[18] Triemstra M, Rosendaal FR, Smit C, Van der Ploeg HM, Briet E. Mortality in patients with hemophilia: changes in a Dutch population from 1986 to 1992 and 1973 to 1986. Ann Intern Med 1995;123:823–7.

[19] Personal communication by Joseph Pugliese, Hemophilia Alliance.

[20] Curran JW, Evatt BL, Lawrence DN. Acquired immunodeficiency syndrome: the past as prologue. Ann Intern Med 1983;98:401–3.

[21] Ricky Ray Hemophilia Relief Act of 1998; Public Law 105-369; November 12, 1998.

[22] Key features of the Affordable Care Act by year. Accessed from: <www.HHS.gov/HealthCare>; May 16, 2014.

[23] Connor JD. Court decision in transfusion-associated AIDS case. Letter from LifeSource; July 15, 1987.

[24] Weinberg PD, Hounshell J, Sherman LA, Godwin J, Ali S, Tomori C, et al. Legal, financial, and public health consequences of HIV contamination of blood and blood products in the 1980s and 1990s. Ann Intern Med 2002;136:312–9.

[25] Dietrich SL. Approach to the clinical management of hemophilia patients at risk for AIDS or the AIDS-related complex. AIDS Center News, Clinical Management Update; April, 1985.

[26] Hilgartner MW. AIDS and hemophilia. N Engl J Med 1987;317:1153–4.

Chapter 2

What Is Hemophilia?

Hemophilia: a medical condition in which the ability of the blood to clot is severely reduced, causing the sufferer to bleed severely from even a slight injury. The condition is typically caused by a hereditary lack of a coagulation factor, most often factor VIII [1].

An infant learning to crawl stops abruptly and shrieks in pain. The mother looks at the child and is unable to see any obvious cause for the distress. She lifts him and he stops crying, but when she puts him down, the crying resumes. She notes that he seems to be favoring one leg; when she moves the leg, he screams with pain. Although not apparent, bleeding has begun in the knee joint, and only after several hours will there be swelling and skin discoloration. In another scenario, blood will suddenly appear in the mouth of a teething infant. On examination, a small bleeding point will be seen on the membrane that attaches the tongue to the floor of the mouth (the frenulum). Pressing gauze against the bleeding site will stop the hemorrhage; however, as soon as the pressure is relaxed, oozing will resume and continue indefinitely. Parents become alarmed and take the child to the pediatrician or a hospital emergency department. Medical personnel see a boy with multiple bruises and might suspect that the child is being abused or neglected, the "battered child syndrome." However, if the possibility of a bleeding disorder is considered, then a simple blood clotting test usually indicates the diagnosis of hemophilia.

Blood within our vessels is fluid because it contains factors that prevent clotting (anticoagulants), and these predominate over the factors required for coagulation (procoagulants). However, if blood is shed from a wound or collected in a glass tube, then it clots rapidly. The clotting time of blood from people with hemophilia is greatly prolonged, suggesting that the composition of their blood is altered. This could be due to a decrease in a procoagulant or an increase in an anticoagulant. The latter hypothesis was championed by Leandro Tocantins (1901–63), who proposed that people with hemophilia bled because their blood contained excessive amounts of an anticoagulant [2]. He supported this theory by experiments showing that blood from a patient with hemophilia prolonged the clotting time of normal blood. What Tocantins did not know was that his patient's blood contained a clotting factor antibody that had developed

Linked by Blood: Hemophilia and AIDS. DOI: http://dx.doi.org/10.1016/B978-0-12-805302-7.00002-1

in response to previous transfusion therapy. At the time Tocantins performed his experiments, he and other doctors were unaware of the existence of these antibodies, which inactivate normal clotting factors. Such antibodies are not found in hemophiliacs who have not undergone transfusions; therefore, their presence does not explain the prolonged clotting times of most people with this condition. It has now been firmly established that the majority of bleeding disorders, including hemophilia, are due to deficiencies or defects in procoagulants and not an excess of anticoagulants.

The clotting factor missing from people with hemophilia was initially called antihemophilic factor or antihemophilic globulin (globulin is the plasma protein fraction that contains the clotting activity); it was discovered in 1937 [3]. Subsequently, several other clotting proteins were identified and given various descriptive names. These names caused considerable confusion because the same factor might be given different names depending on the place and circumstances of its discovery. Eventually, in 1977, the clotting proteins were assigned Roman numerals [4], and the clotting proteins affected in hemophilia were designated factor VIII (missing in approximately 80% of hemophiliacs) and factor IX (decreased or absent in 20%). Factor IX and most other clotting factors are made by liver cells, but factor VIII is produced by the cells that line the walls of blood vessels [5,6]; in this location, it is exactly where it is needed to stanch bleeding.

The incidence of hemophilia A (factor VIII deficiency) is 1 in 10,000, and that of hemophilia B (factor IX deficiency) is 1 in 60,000 [7]. There are historical records of an entity resembling hemophilia dating back to the second century [8], and the disorder has been described in every race and ethnic group.[1] Genes located on the X chromosome code for the production of these coagulation proteins; gene mutations can give rise to deficient or defective clotting factors. Women have two X chromosomes; if genes on only one are mutated, then the unaffected one can provide sufficient clotting factor to prevent excessive bleeding. These women are known as carriers (of the mutant gene), and they usually have lower levels of hemophilic factors than other women. Because males have only one X chromosome, mutations in the gene on this chromosome can greatly impair clotting factor formation and can be associated with a lifelong bleeding tendency. More than 2100 unique mutations in the factor VIII gene and 1100 in the factor IX gene have been identified [9]. Approximately half of all families that have a boy with hemophilia cannot recall any relatives with this disorder, and many women learn that they are carriers only when they have an affected infant.

The severity of hemophilia is related to the specific mutation or mutations present, and these determine the amount of clotting factor in the blood. In persons with severe hemophilia (who have less than 1% of the normal level of the clotting factor), bleeding occurs without provocation and as often as once per

---

1. For example, hemophilia was described in members of the Bantu people of South Africa by Gomperts E, Lurie A, Greig HBW, Katz J, Metz J. Haemophilia in the Bantu: report of 67 cases. S Afr Med J 1969;934–5.

week. More mildly affected individuals might have 5% or 10% of the normal amount of clotting protein and bleed only when injured or undergoing surgery.

Bleeding occurs most frequently in the knees and elbows, but the ankles, shoulders, and hips are also affected. Muscle hemorrhages and bleeding into the brain and other organs is less common but more dangerous. The pain of a joint bleed can be excruciating, and bed rest and ice compresses provide only partial relief. Blood leaking into the joint space causes inflammation and the release of enzymes that destroy the fragile cartilage. The joint lining becomes swollen, thickened, and scarred. The eventual outcome is a frozen, unworkable joint. Thus, hemophilia is a disease that cripples, and recurrent bleeds and loss of mobility result in chronic disability. Figs. 2.1–2.3 display muscle and joint hemorrhages in hemophiliacs.

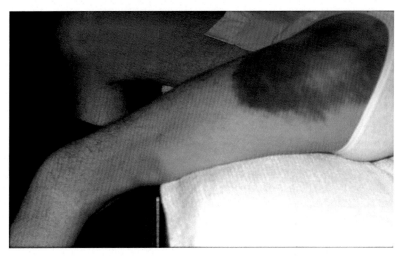

FIGURE 2.1   This young man injured his thigh in a bicycle accident. The small bruise that appeared initially quickly expanded to become this large hematoma. *Photograph courtesy of the author.*

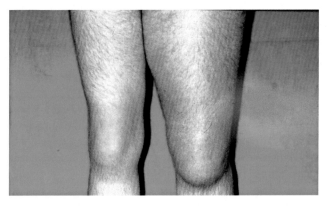

FIGURE 2.2   This left thigh muscle hemorrhage began spontaneously, and the blood eventually made its way down into the knee joint. *Photograph courtesy of the author.*

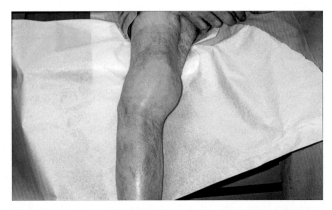

FIGURE 2.3   As a consequence of repeated hemorrhages into the knee, the joint has become deformed and dysfunctional, and the leg muscles have atrophied. *Photograph courtesy of the author.*

The psychological impact of this condition cannot be overestimated. For the affected child, the unpredictability of the painful bleeds, the frequent visits to hospital emergency departments, and the disruption of ordinary activities promote an inward focus and a craving for pain relief and constant support. Attendance at nursery school is intermittent at best, and play dates with other children are infrequent and often need to be canceled because of bleeding. School attendance is spotty and education becomes discontinuous. Participation in after-school activities and sports is negligible. The affected boy is viewed as a weakling and shunned by his peers, and his social life is constricted. Mothers become obsessed with the care of their hemophilic sons and attempt to protect them from even the slightest injury; they pad their surroundings and supervise every step. These mothers become reluctant to permit any independent activity; for example, the hemophilic sons of one woman explained that they had "no" disease, because they were told "no" every time they wanted to do something. A bleeding episode is interpreted as a failure to safeguard the child's health. Guilt is also engendered by the mother's recognition that her mutated gene was transmitted to her son. The constant concern about her son's health impacts her relationship with her husband. Planned social events, evenings out, and conjugal relations are interrupted by the unpredictable nature of the child's hemorrhages. Fathers become wary of even touching their sons; rough-housing and most sports are strictly avoided. Siblings experience a lack of attention from parents preoccupied with the affected child and often become hostile. Then, they feel guilt when their brother has a painful bleed. Although boys with hemophilia often suffer from considerable psychological stress, there does not appear to be an association between the frequency of bleeding episodes and measured psychological factors [10].

The most striking example of the devastating effects of hemophilia is illustrated by the story of Alexis, the heir to the Russian throne at the turn of the

19th century. Alexandra, his mother and a daughter of Queen Victoria, was a carrier of the mutant hemophilia gene. She had four healthy daughters, and then Alexis was born with factor IX hemophilia [11]. He experienced repeated painful hemorrhages. Because none of the then-available remedies could prevent bleeding and relieve pain, Alexandra consulted the monk, Rasputin, who was an accomplished hypnotist. Using hypnosis, he was able to provide some relief to the boy, and his grateful mother established Rasputin as an accepted member of the royal entourage [12]. Unfortunately, hypnosis could not prevent recurrent episodes of joint bleeding and cartilage destruction, and the Prince became severely crippled. It has been suggested that these circumstances—a crippled heir to the Throne, anxiety about preserving his protective environment, and Rasputin's influence on the Royal Family—all encouraged the Bolshevik revolution that led to the downfall of the Monarchy [13].

The famous Greek physician, Hippocrates, recommended that hemorrhages be treated by the application of cold to the area around a bleeding site. Two thousand years later, investigators from the Mayo Clinic confirmed that blood loss from skin incisions in hemophiliacs was decreased if the surrounding skin was chilled, but it was enhanced if cold was applied to the wound itself [14]. Cold promotes constriction of blood vessels, which reduces the amount of blood reaching the site of bleeding, but the blood in the wound actually clots best at body temperature. Other measures used for the management of painful hemorrhages were purging, hot irons, mercury, silver nitrate, and tannins [15]; in the early part of the past century, peanuts were thought to hasten blood clotting, and some hemophiliacs were convinced they were beneficial, but scientific proof of their efficacy was never demonstrated.

When physicians were unable to control bleeding or relieve pain, hemophiliacs and their families often resorted to alternative remedies. In a recent essay, a writer from South America described how her mother tried to force her hemophilic brother to drink "miracle water"; the boy recognized that this was a quack remedy and refused it [16]. This mother also applied poultices of crushed volcanic rock, burned lamb's wool, or volcanic oil to relieve joint hemorrhages, trying in vain to find a cure for her son's illness.

As early as 1840, it was reported that bleeding could be controlled by injecting blood from a healthy person into the vein of a person with hemophilia [17]. This was one of the first accounts of a successful transfusion. Chapter 3, describes how the safety of transfusion was improved by blood typing, and the discovery that certain salts, such as sodium citrate, could prevent the clotting of blood after it was removed from the body [18]. These advances enabled the development of blood banks that could collect and maintain blood in stable condition for several weeks. Initially, blood was kept in bottles and transfused through rubber tubing and large-bore needles inserted into a vein. Transfusing the blood was a tedious and time-consuming process, and often the blood would form clots in the bottle or tubing. A great improvement was the development of plastic bags and tubing. This enabled processing of blood into red cell and plasma components; centrifugation of a pint of blood produces a bag of packed red blood cells and approximately half a pint of clear, yellow plasma. It is easier to store and infuse the bags of red blood cells, and the plasma can be frozen and stored for up to 1 year as fresh-frozen plasma (FFP). After thawing, the FFP is administered by intravenous infusion.

The treatment of hemophilic bleeding requires augmenting the level of the clotting factor in the patient's blood. Raising the clotting factor level to 15–20% of normal can prevent spontaneous bleeding, but levels of 50–100% are necessary to control a major hemorrhage. The amount of FFP needed to achieve these levels depends on the person's size; larger individuals require more plasma. To stop bleeding from the tongue of a baby, 1 or 2 ounces might be sufficient, whereas controlling major bleeding in a 200-pound man might require several pints of FFP.

Bleeding in people with hemophilia can occur unpredictably at any time of the day or night. In the era prior to the development of clotting factor concentrates, most hemophiliacs with a hemorrhage were brought to a 24-hour facility such as a hospital emergency department. This was because hospitals had access to blood products, and hospital personnel were qualified to administer the FFP needed to control bleeding. In the emergency department, the hemophiliac individual had to wait to be seen by a physician, and there was an additional wait for the frozen plasma to thaw. Medical personnel often had difficulty accessing the tiny and frequently used veins of small children and adults with hemophilia; successful needle insertion usually required multiple attempts, causing a great deal of pain and bruising. Finally, there was the long wait for the slow dripping of the plasma into the vein, which eventually stopped the bleeding. Not infrequently, the whole family spent the night in the emergency department, and this scenario was often repeated several times every month. Family life revolved around the continuous need for FFP and hospital visits; this became the focus of all family activities. The financial costs were considerable; they included physician and laboratory fees and charges for the FFP, transfusion equipment, and the emergency department.

Although volunteers give blood without receiving financial compensation, the blood is not free to those who need it. Blood banks generally affix a price based on the costs of collecting, typing, storing, and dispensing the blood or FFP. And, of course, the supply is limited by the number of donors available to provide the blood. In the past, those with severe hemophilia generally required transfusions three or four times per month to stop minor bleeds; if there was a larger hemorrhage, then FFP might be needed on a daily basis for up to 3 weeks. How could this terrible burden of emotional turmoil, time, and expense be borne by an average middle-class family?

Families recognized that they needed to mobilize members of their community to donate blood so that FFP would always be available. They learned to negotiate with blood banks so that they would receive a credit for each unit donated, and these blood credits could be used to cover the costs of the FFP.

Families sought assistance from their churches, schools, hospitals, Red Cross, or community sources to organize and run blood drives. A family in my practice had three hemophilic sons; they were able to collect as many as 860 units of plasma annually through frequent blood drives. Their parents became very knowledgeable about the treatment of hemophilia, and when they read that a 6-month-old infant with hemophilia had been found abandoned on Christmas day, they adopted the baby. When interviewed, they said that "…we felt we couldn't let a child be condemned to a life in the hospital when we had the experience to help him" [19].

Hemophilic families banded together and organized local chapters and national groups to provide mutual assistance and support. The National Hemophilia Foundation (NHF) was formed in 1948; by 1982, 50–70% of hemophiliacs in the United States were members of local chapters or were enrolled in regional Hemophilia Treatment Centers. The mission of the NHF was to seek better treatment and prevent complications of bleeding disorders through education, advocacy, and research. To provide its membership with expert opinion, the NHF selected physicians who were knowledgeable in the treatment of hemophilia to serve on its Medical and Scientific Advisory Council (MASAC). This council advised the NHF about advances in the management of hemophilia and made recommendations for communication to the membership.

The use of FFP to treat hemophilia is unsatisfactory in many respects. Quantitative considerations show that achieving a substantial increase in clotting factor levels in adults when treating with normal plasma is challenging because the half pint of FFP being infused is diluted by the 5 quarts of blood already circulating in a normal adult. Each additional bag of infused FFP further increases the amount of fluid in the circulation and only incrementally raises clotting factor levels; therefore, even several bags might not increase clotting factor concentrations sufficiently to control major bleeding. The logistics of storing, thawing, and administering such large amounts of FFP can be daunting, especially in emergency situations, as the following case study illustrates.

In the late 1960s, a 26-year-old man with hemophilia was admitted to our hospital because earlier that day he had slipped on a newly waxed kitchen floor. He immediately noticed pain in his right calf, and soon thereafter he noticed swelling of his right lower leg. Soon, a large bruise appeared by his ankle and his foot began to swell and throb. On examination, the right calf was distended with blood and the foot was dusky in color and cool to the touch. A compartment syndrome was suspected; this occurs when blood is trapped within a muscle compartment and compresses the nerves and vascular structures located therein. He was immediately taken to the operating room for decompression surgery to relieve the pressure on the nerves and blood vessels; this required flaying open the muscle and packing the wound with gauze. To control bleeding, he was given 10 bags of FFP; this dose was chosen because he weighed more than 200 pounds and it was calculated that at least that amount would be required to provide a modest increase in his level of clotting factor. Despite this treatment and additional units of plasma, oozing from the cut surface of the muscle persisted; the presence of blood in the tissues contributed to the whole surgical site becoming infected. Antibiotics were administered without effect, and it was deemed necessary to amputate the leg below the knee. However, the new surgical wound bled and became infected, and the amputation had to be advanced to the mid-thigh area despite the administration of dozens of units of blood and FFP as well as antibiotics and pain medications. Three months after the original injury, the wound was still gaping and infected, and it extended almost to his hip; the dressings covering this huge area were constantly soaked with blood and purulent material. It was not considered feasible to cover the wound with a full-thickness skin graft taken from some other area of his body because there would be bleeding from that site. Our plastic surgeon very cleverly improvised by collecting multiple small, postage stamp–sized grafts from several skin areas and using them to cover the wound. These grafts stanched the bleeding and gradually coalesced to completely heal the wound, enabling our patient to be discharged from the hospital. Although he contracted hepatitis from the many transfusions he was given, his condition eventually stabilized and he learned to use crutches and drive a car. He raised a daughter who became a nurse specializing in the treatment of people with hemophilia. During the early 1980s, our patient became infected with human immunodeficiency virus (HIV), but he had a good response to antiviral medications. He died at the age of 67, more than 40 years after what originally was considered to be a fatal hemorrhage. This case history illustrates that even a trivial accident in a hemophiliac can induce life-threatening hemorrhage, that FFP is an imperfect treatment for bleeding, and that courageous individuals can rebuild lives shattered by serious illness or injury.

A major breakthrough in hemophilia care occurred in 1965, when a simple method for obtaining a concentrated solution of factor VIII was described [20]. The procedure developed by Pool and Shannon consisted of centrifuging the blood to separate the plasma from the red blood cells, placing the plasma in a plastic bag, and rapidly freezing it. The frozen plasma was then thawed in a refrigerator. A milky white precipitate formed in the plasma and coated the bag's inner surface. The clear plasma was removed and the precipitate, called cryoprecipitate, was refrozen and stored in the freezer. When needed for transfusion, the addition of a small amount of warm plasma dissolved the precipitate. More than half the factor VIII originally present in the frozen plasma was now concentrated within the resulting viscous yellow liquid, and the volume had shrunk from a half a pint to two tablespoons. The contents of several bags could be pooled into a single bag (Fig. 2.4), and the material could be infused intravenously [21].

Cryoprecipitate has many advantages over FFP. Its production requires only a centrifuge, freezer, and refrigerator—equipment that is available in every blood bank. The red blood cells and remaining plasma can be set aside for patients needing those products. Storage of the cryoprecipitate requires much

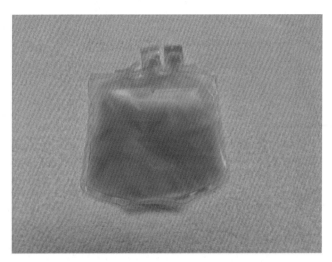

FIGURE 2.4 Cryoprecipitate from five donors pooled in one bag. *Photograph courtesy of the author.*

less freezer space than whole plasma. The material can be thawed and reconstituted more rapidly than FFP and requires much less time for infusion. In fact, instead of a slow intravenous drip, the dissolved cryoprecipitate can be drawn into a syringe and injected through an inline filter directly into a vein.

The ease of storage and administration of cryoprecipitate led to the introduction of home care by Rabiner and Telfer [22]. Whereas previous treatment had been administered mainly in hospital emergency departments or doctor's offices, it now became possible to teach hemophiliacs and family members how to store and reconstitute the clotting factor at home. They were instructed in the technique of vein puncture and intravenous infusion. A designated physician or nurse was called whenever bleeding was suspected and made the decision about whether the patient should come to the hospital for examination or could be treated at home. After a brief period of training, most patients were able to be treated at home.

It even became possible to use the materials to prevent bleeding; for example, the cryoprecipitate could be administered prior to participation in contact sports [23]. The benefits were enormous; hemophiliacs became enrolled in regular school programs, including gym periods, and participated in non-contact sports.[2] They were absent less often from school or work. And, for the first time, boys with hemophilia could have a summer camp experience, learn to self-infuse

---

2. Recommended sports include swimming, table tennis, fishing, dance, badminton, sailing, golf, bowling, cycling; sports to be avoided are boxing, football, karate, wrestling, judo, motorcycling, hockey, skateboarding (Jones PM, Buzzard BM. Hemophilia and sport. In: Forbes CD, Aledort L, Madhok R, editors. Hemophilia. New York, NY: Chapman & Hill; 1997. p. 136).

clotting factor, and become independent. For example, at Camp Warren Jyrch in Illinois, 21 boys aged 7–15 years successfully performed venipunctures and 10 boys began to regularly infuse themselves with clotting factor (see the frontispiece) [24]. They learned that hemophilia was not limited to persons of a single race or ethnicity, and they became more accepting of their disabilities. The camp experience offered a unique opportunity for education, fostered self-reliance, and increased self-esteem, and it was probably best summed up by the statement of one mother: "I sent a boy to camp, I got back a man!" [24].

Pharmaceutical companies recognized that plasma and cryoprecipitate could be used as the starting point for manufacturing clotting proteins. Because only the plasma portion of the blood was needed for preparing clotting factor concentrates, donors could be bled and the red blood cells could be returned to them. This procedure is called plasmapheresis, and it permitted plasma donations from individuals on a weekly basis. To ensure a consistent source of plasma, manufacturers established plasma collection centers, usually in the poorer neighborhoods of large cities where there were unemployed persons who had the time to donate their plasma and were eager to receive the small fees paid for each donation. Plasma collected from hundreds of donors was pooled and the clotting protein was extracted, concentrated, and freeze-dried in vials. This material could be stored for up to 1 year and then reconstituted. The final product was shipped to blood banks and pharmacies throughout the world. The vials of concentrate were easy for patients to store and reconstitute, and they could be kept in lockers at school or at the workplace. Companies provided travel kits containing needles and syringes as well as the vials of concentrate so that travel over long distances could be accomplished knowing that treatment for an unexpected hemorrhage was readily available. By the end of the 1970s, many persons with hemophilia were able to have a conventional lifestyle and avoid most of the devastating bleeding that led to crippling and death during earlier periods. The blood product had become central and essential to the life of every person with hemophilia.

Persons with hemophilia and family members obtained clotting factor concentrates from their hemophilia centers, hospital blood banks, or pharmacies. For example, in 1982, a group of New York hemophilia center physicians formed a consortium to purchase and distribute commercial clotting factor concentrates [25]. By purchasing large amounts of product, they were able to secure below-market prices from the manufacturers. The concentrate was prescribed to patients, whose insurers were billed the cost of the product plus a surcharge of 12.5%. This arrangement provided funds for home delivery of the concentrate and injection supplies, as well as for the administrative expenses of the consortium. The ready access to clotting factor concentrate greatly simplified the treatment of hemophilia.

More effective therapy enabled men with hemophilia to pursue rewarding careers. Because they were barred from most competitive sports during their childhood, many developed an intense desire to succeed during adulthood.

They achieved success in fields such as insurance, accounting, broadcasting, and photography, among others. Several were attracted to careers in health care and contributed to knowledge about bleeding disorders. The wife of one man recalled "My late husband…never let his hemophilia stop him and he tried to live as 'normal' a life as he could in spite of the challenges" [26]. She noted that he was an ardent fisherman because this activity was gentle to his joints and enabled him to interact with others who loved the sport.

In 1975, the US Congress established the Hemophilia Diagnostic and Treatment Center Program, organizing hemophilia management around 24 Comprehensive Care Centers [27]. Each center included physicians, nurses, orthopedists, dentists, psychologists, and social workers with an interest and expertise in hemophilia care. The objectives were to prevent or contain bleeding, restore joint function, and provide counseling about educational, employment, and financial problems. Within a decade of their inception, these centers were serving approximately half of the nation's hemophiliacs and had proven to be so successful that they were awarded $3.5 million to expand the comprehensive care concept [28]. They continue to provide complete health care services for two-thirds of the nation's hemophiliacs.

By the late 1970s, the majority of hemophiliacs were receiving clotting factor concentrates on a regular basis. The median age at death increased from 33 to 55 years during the years from 1968 to 1979 [29], and the median life expectancy at age 1 year of patients with severe hemophilia increased from 39.7 years during 1941–60 to 60.5 years during 1971–80 [30]. The causes of death, in addition to fatal bleeding, were liver disease secondary to hepatitis [31] and pneumonia. In addition, the use of clotting factor concentrates resulted in a profound decrease in deaths from bleeding into the head, which was previously one of the most terrible afflictions of hemophiliacs. Between 1957 and 1968, such bleeding accounted for one-third of deaths of young hemophiliacs in Sweden, but there were none from 1968 to 1980 [32].

The impact of clotting factor concentrates on the hemophilia community cannot be overstated [33]. Before the introduction of this therapy, patients and families were helpless in the face of ongoing bleeding. When clotting factor concentrates became available, patient lives became immeasurably improved [34]. Although the treatment required intravenous infusions, mothers and fathers learned to access the veins of boys as young as 4 years of age and infuse them with the concentrated clotting factor. This treatment rapidly stopped bleeding, controlled pain, and enabled early resumption of activities. It even became possible to safely perform surgical procedures, such as the replacement of joints damaged by previous hemorrhages. Dr Margaret Hilgartner, director of one of the hemophilia treatment centers, succinctly summed up the prevailing philosophy, "When in doubt, treat," by which she meant infuse clotting concentrate whenever a hemorrhage was suspected.

Although clotting factor concentrates were effective, they were associated with a number of problems. First, they needed to be infused intravenously, which meant that a needle had to be placed in a vein and remain securely attached to the arm while the clotting factor was infusing. Often, the needle might pierce the vein wall or become dislodged, and the intravenous fluid would enter the tissues around the vein, causing pain, swelling, and redness. Such an occurrence required the needle to be removed and placed in a new site, usually in the other arm. Over time, crude large-bore metal needles were replaced by thin plastic cannulas, and extravasations of fluid from the vein became less common. Second, the infused material might provoke an allergic reaction, usually consisting of itching, swelling, and hives. Treatment would have to be discontinued and an antihistamine or similar medicine would have to be administered. Over time, these adverse events became less frequent and severe as manufacturers removed more impurities from the concentrates.

A far more intractable problem with clotting factor replacement therapy was the development of inhibitors. Inhibitors are antibodies directed against coagulation proteins; they are formed because the immune system interprets the infused proteins as foreign invaders and attempts to destroy them. Clotting factors in the concentrate are inactivated by the antibodies, and the hemorrhage that is the target of treatment continues unabated, to the dismay of the patient and his physicians. The appearance of the inhibitor is usually quite subtle; bleeds that previously had been easily controlled with a small dose of clotting factor require progressively more and more concentrate until they become completely refractory to even the largest doses. In the early years of hemophilia treatment, the appearance of inhibitors was infrequent, but with the advent of more intense exposure to highly purified concentrates they are reported in as many as 30% of all hemophiliacs. Currently, it is recommended that testing for inhibitors should be performed annually so that alternative treatments for bleeding can be implemented. The development of inhibitors has become one of the most serious problems currently encountered in the management of hemophilia.

There was another, potentially deadly, concern with clotting factor concentrates. Because these products were prepared from pools of hundreds of donors, there was the possibility that if even one donor in the pool harbored an infection, the infecting agent could be transmitted to all the recipients of the blood product containing that donor's blood. For example, if a donor was infected with the hepatitis virus, then concentrate prepared from that donor's blood could induce hepatitis in everyone receiving that product. Transmission of hepatitis viruses became widespread; as early as 1971, 17% of hemophiliacs who underwent transfusions were found to be positive for hepatitis-associated antigen [35]; by 1982, as the use of blood products increased, the incidence of hepatitis B reached 5.4%, and that of non-A, non-B hepatitis (hepatitis C) reached 27% [36]. Liver cirrhosis and chronic active hepatitis were reported in 15% and 7% of hemophiliacs, respectively [37]. Dr Oscar Ratnoff, a blood coagulation expert and distinguished scientist, wrote that in his hometown of Cleveland, OH, the concentrate manufacturer's plasma collection facility was located on skid row [38]. Many of the donors were drug abusers infected by hepatitis viruses. Ratnoff recognized that commercial clotting factor concentrates could transmit infection and therefore used only cryoprecipitate donated by patients' families and friends, but he was the exception. Although most doctors and patients were aware that blood and blood products could be contaminated with disease-causing viruses, transfusion therapy with high potency concentrates had become central to the management of hemophilia. This set the stage for the widespread transmission of the human immunodeficiency virus (HIV), the cause of acquired immunodeficiency syndrome, to persons with hemophilia.

KEY POINTS

- Hemophilia is a genetic disorder due to mutations in the genes for clotting factor VIII or factor IX; these genes are located on the X chromosome. To date, more than 2100 unique mutations in the factor VIII gene and 1100 in the factor IX gene have been described.
- Hemophilia may be mild, moderate, or severe, corresponding to the level of the clotting factor in the blood (>5%, 1–5%, <1%, respectively, compared to 50–150% in healthy persons).
- The incidence of hemophilia is 1 in 5000 males, of whom 80% have factor VIII deficiency and 20% have factor IX deficiency.
- Hemorrhages first appear when babies begin teething and crawling. Toddlers have bleeding into knees, ankles, and elbows; this bleeding produces pain, swelling, joint lining (synovial) thickening, and scarring (fibrosis). The end result of untreated disease is crippling.
- The advent of cryoprecipitate and clotting factor concentrates in the 1970s immeasurably improved the lives of hemophiliacs.
- At the time of the AIDS epidemic, therapeutic products for hemophiliacs were fresh frozen plasma (FFP), cryoprecipitate made from individual plasma donations, and commercially prepared clotting factor concentrates made from pooled plasma donations.
- The life expectancy of hemophiliacs before the availability of blood products was 39.7 years. By 1980, when the use of cryoprecipitate and concentrates became widespread, it was 60.5 years.
- The major complication of treatment was viral hepatitis transmitted by blood products.

REFERENCES

[1] Oxford English Dictionary [accessed 05.02.14].
[2] Tocantins L. Hemophilic syndromes and hemophilia. Blood 1954;9:281–5.
[3] Patek Jr. AJ, Taylor FHL. Hemophilia. II. Some properties of a substance obtained from normal plasma effective in accelerating the clotting of hemophilic blood. J Clin Invest 1937;16:113–24.

[4] Jackson CM. Recommended nomenclature for blood clotting zymogens and zymogen activation products of the International Committee on Thrombosis and Haemostasis. Thromb Haemost 1977;38:567–77.

[5] Fahs SA, Hille MT, Shi Q, Weller H, Montgomery RR. A conditional knockout mouse model reveals endothelial cells as the predominant and possibly exclusive source of plasma factor VIII. Blood 2014;123:3706–13.

[6] Everett LA, Cleuren ACA, Khoriaty RN, Ginsburg D. Murine coagulation factor VIII is synthesized in endothelial cells. Blood 2014;123:3697–705.

[7] Mannucci PM, Tuddenham EGD. The hemophilias-from royal genes to gene therapy. N Engl J Med 2001;344:1773–9.

[8] Ratnoff OD. Why do people bleed? Wintrobe MM, editor. Blood, pure and eloquent. New York, NY: McGraw-Hill; 1980, p. 625.

[9] Peyvandi F, Kunicki T, Lillicrap D. Genetic sequence analysis of inherited bleeding diseases. Blood 2013;122:3423–31.

[10] Cochran CD, Ahles TA, Weiss AE. Psychological factors and bleeding frequency in hemophilia: lack of association. Am J Pediatr Hematol Oncol 1987;9:136–9.

[11] Rogaev EI, Grigorenko AP, Faskhutdinova G, Kittler EL, Moliaka YK. Genotype analysis identifies the cause of the "Royal Disease". Science 2009;326:817.

[12] Lucas ON. The use of hypnosis in hemophilia dental care. Ann NY Acad Sci 1975;240:263–6.

[13] Massie RK. Nicholas and Alexandra. New York, NY: Atheneum; 1967, p.595.

[14] Sutor AH, Bowie EJW, Owen Jr. CA. Effect of cold on bleeding: Hippocrates vindicated. Lancet 1970;ii:1084.

[15] Legg JW. Treatise on haemophilia. London: HK Lewis; 1872, p.158.

[16] Mulligan MH. Taking the cure. NY Times 2014; July 13:50.

[17] Lane S. Haemorrhagic diathesis-successful transfusion of blood. Lancet 1840;1:185–8.

[18] Diamond LK. A history of blood transfusion Wintrobe MM, editor. Blood, pure and eloquent. New York, NY: McGraw-Hill; 1980.

[19] Kloss J. Room for 1 more in merciful family. Chicago Daily News 1970; March 11.

[20] Pool JG, Shannon AE. Production of high-potency concentrates of antihemophilic globulin in a closed-bag system. New Engl J Med 1965;273:1443–7.

[21] Chicago Blood Donor Service, Inc. Preparation of antihemophilic globulin-rich cryoprecipitate. Newsletter 1967:1–4.

[22] Rabiner SF, Telfer MC. Home transfusion for patients with hemophilia A. N Engl J Med 1970;283:1011–5.

[23] Lazerson J. The prophylactic approach to hemophilia A. Hosp Pract 1971:99–109.

[24a] Seeler RI, Ashenhurst JB, Miller J. A summer camp for boys with hemophilia. J Pediatr 1975;87:758–9.

[24b] Seeler RA, Ashenhurst JB, Langehennig PL. Behavioral benefits in hemophilia as noted at a special summer camp. Clin Pediatr (Phila) 1977;16:525–9.

[25] Aledort LM, Lipton RA, Hilgartner M. A consortium for purchase of blood products directed by physicians. Ann Intern Med 1988;108:754–6.

[26] The FactorNet; Spring 2014, p. 7.

[27a] Smith PS, Keyes NC, Forman EN. Socioeconomic evaluation of a state-funded comprehensive hemophilia-care program. N Engl J Med 1982;306:575–9.

[27b] Aledort LM. Lessons from hemophilia. N Engl J Med 1982;306:607–8.

[28] National Hemophilia Foundation. Treatment center funding increase. Hemophilia Information Exchange; December 1984.

[29] Aronson DL. Cause of death in hemophilia A patients in the United States from 1968 to 1979. Am J Hematol 1988;27:7–12.

[30] Jones PK, Ratnoff OD. The changing prognosis of classic hemophilia (factor VIII "deficiency"). Ann Intern Med 1991;114:641–8.

[31] Hay CRM, Preston FE, Triger DR, Underwood JCE. Progressive liver disease in haemophilia: an understated problem? Lancet 1985;i:1495–8.

[32] Larsson SA, Wiechel B. Deaths in Swedish hemophiliacs, 1957–1980. Acta Med Scand 1983;214:199–206.

[33] Snider AJ. New hope for hemophiliacs. Chicago Daily News 1968;Saturday, February 17.

[34] Hilgartner M. Therapeutic advances in hemophilia and von Willebrand's disease, p. 610–3. In: Gralnick HR (moderator): Factor VIII. Ann Intern Med 1977;86:598–616.

[35] Seeler RA, Mufson MA. Development and persistence of antibody to hepatitis-associated (Australia) antigen in patients with hemophilia. J Infect Dis 1971;123:279–83.

[36] Rickard KA, Dority P, Campbell J, Batey RG, Johnson S, Hodgson J. Hepatitis and haemophilia therapy in Australia. Lancet 1982;ii:146–8.

[37] Aledort LM, Levine PH, Hilgartner M, Blatt P, Spero JA, Goldberg JD, et al. A study of liver biopsies and liver disease among hemophiliacs. Blood 1985;66:367–72.

[38] Ratnoff OD. Some complications of the therapy of classic hemophilia. J Lab Clin Med 1984;103:653–9.

Chapter 3

Blood: Vital but Potentially Dangerous

The life of the flesh is in the blood.

Leviticus 17:11

Blood has a powerful mystique, eliciting respect but instilling fear. People admire a full-blooded animal but cringe at the sight of blood. Since ancient times, blood has been equated with life. The Bible enjoins one to refrain from "eating" blood because "I have given it to you upon the altar to make atonement for your souls; for it is the blood that maketh atonement by reason of the life" [1]. Blood has always been a substance both revered and reviled. One speaks of "life blood" or "precious drops of blood" when describing its life-sustaining qualities, but uses the term "bloody" as a curse word. Blood is used as an adjective in describing people, objects, and activities: "blood brother," "blood money," and "blood sport," but it is also used to connote something more ominous, as in "blood bath," "blood feud," and "blood-thirsty." Blood may also be "bad" when relationships have soured (bad blood) or "cold" when referring to someone lacking compassion or acting without remorse: "a cold-blooded murderer" or "slain in cold blood." Aristocrats are said to have "blue" or "royal" blood; however, in Shakespeare's time, Macbeth says of the King's corpse: "Here lay Duncan, /His silver skin laced with his golden blood" [2]. However, Lady Macbeth is consumed with guilt and cannot wash this blood from her hands: "Out, damned spot! Out, I say!" But the blood stain resists removal: "What, will these hands ne'er be clean?" [3]. Thus, blood is often given mystic powers, and the many terms used to describe it reflect our special relationship with this life-sustaining liquid that plays such an important role in our daily existence.

Blood has figured prominently in religious ceremonies. For example, in Roman times, a priest would sacrifice a bull and shower the blood on a citizen seeking spiritual rebirth. The blood was considered to have rejuvenating power [4]. Christians receiving the Eucharist or Holy Communion remember

Linked by Blood: Hemophilia and AIDS. DOI: http://dx.doi.org/10.1016/B978-0-12-805302-7.00003-3

the instruction of Jesus at the Last Supper, when He gave his disciples bread, saying "This is my body," and wine, saying "This is my blood." During the festive dinner (Seder) on Passover, Jews recall the plagues that God inflicted on the Egyptians. Those attending the ceremony recite aloud the name of each of the 10 plagues, spilling a drop of wine as each is intoned. The first plague was "blood" ("dam" in Hebrew). These rituals demonstrate that blood has great symbolic significance; in addition, it can arouse strong emotions. For example, some people develop a pathologic fear of blood; the mere sight or even the thought of seeing blood induces fainting. The main protagonist in the contemporary British television comedy series, "Doc Martin," has to relinquish his career as a surgeon because of a violent reaction (vomiting) at the sight of blood.

Blood is not a homogeneous liquid; it contains various cellular elements. These were initially described by Antonj van Leeuwenhoek in 1674 [5]. He wrote: "The Blood is composed of exceeding small particles, named globules, which, in most animals are of a red color, swimming in a liquor, called by physicians, the serum…" [5]. Today, we know that the average person weighing 150 pounds has approximately 5 quarts of blood containing approximately 2 quarts of cells comprising 10^{13} (more than 20 trillion) individual cells. This cellular compartment contains red cells, white cells, and platelets; all are progeny of parent or stem cells that reside in the bone marrow. The red cells provide the tissues with oxygen and remove carbon dioxide, the white cells fight infection, and the platelets assist in blood coagulation. Individual red cells circulate for 120 days and platelets circulate for 10 days, but most white cells have a lifespan of only a few hours.

During febrile illnesses, a thick green layer of white cells and proteins appears on the surface of blood allowed to clot in a test tube. This was called phlegm by ancient philosophers and was thought to cause disease; removing the blood was seen as a way of ridding the body of this noxious material [6]. Blood-letting became the preferred treatment for many disorders well into the 19th century. The practice was accompanied by purging with emetics; there is no doubt that the major loss of bodily fluids from these "treatments" resulted in a decrease in blood pressure (shock) and the death of many sufferers. Most remarkably, blood-letting was even applied to control hemorrhages in persons with hemophilia! [7]. In his 1872 *Treatise on Haemophilia*, the English physician, Wickham Legg, warned that "the terrible danger attending its use (blood-letting) must deter every careful practitioner from such an idea." He stated further that "death has very frequently followed its employment, and the danger of this more than counterbalances any probable good" [8]. He and others advocated purging; this practice probably induced dehydration and decreased blood pressure, possibly benefiting a bleeding person by slowing the rate of blood loss.

Anemia was one of the first clinical disorders directly linked to blood. For hundreds of years, a condition called chlorosis had been recognized in young women, and it often had a fatal outcome. These women usually had a history of heavy menstrual periods and/or multiple pregnancies, and their skin was a sickly green color. They were probably very anemic; a study by Gabriel Andral in 1845 reported that their red blood cells were remarkably small [9]. A decrease in the size of the red cells is characteristic of iron deficiency anemia. Iron is essential for red cell production, and most of the body's iron resides in the red blood cells. This iron is lost when there is excessive menstrual bleeding or blood loss at the time of childbirth. During the 19th century, physicians confirmed that chlorotic blood contained less than the normal amount of iron, and they found that iron preparations were an effective treatment [10]. A method for dying red cells so that the small cells of iron deficiency could be differentiated from the large cells found in vitamin B_{12} deficiency (the cause of pernicious anemia) was subsequently developed by the Nobel Prize–winning physician and scientist, Paul Ehrlich (1854–1915) [11].

Although iron salts are effective in treating anemia due to chronic blood loss, acute bleeding requires a more rapid method for restoring blood volume. This is accomplished by blood transfusion, which was first attempted in dogs as early as 1666 [12]. The success of transfusion in animals was followed by the infusion of animal blood into humans, with often fatal outcomes, and led to a moratorium on further transfusions. However, in 1818, Blundell reported successful transfusions in four women with postpartum hemorrhages by using blood from their husbands [13]. However, subsequent experiences with transfusion were dismal and the practice was abandoned until the 20th century.

In 1901, Karl Landsteiner, a Viennese physician, discovered the existence of blood groups [14]. He separated serum (the liquid appearing after blood has clotted) from the red cells and showed that sera from some individuals would cause clumping (agglutination) of other persons' red cells, and vice versa. The first group was said to have group A cells and the second group was said to have group B cells. In addition, there were instances in which the cells would not be agglutinated by sera from either group A or group B, and he called these group C (now called group O) cells. Subsequently, a fourth group, AB, was delineated in which the cells are agglutinated by sera from groups A and B. These were enormously important discoveries, and Landsteiner was awarded the Nobel Prize. Since that time, many other blood groups (Rh, Duffy, Kell) were identified and characterized.

The red cell agglutination that Landsteiner described occurs whenever there is a mismatch between the cells of the donor and those of the recipient. It is due to the formation of antibodies by the recipient that recognize that the donor red cells are nonself or foreign. The antibodies in the serum attach to the red cells and cause them to agglutinate. This is an example of an immune response that has evolved to protect individuals from potentially dangerous nonself organisms such as viruses and bacteria. Humans are immersed in a world of microbes that are trying to survive by feasting on the body's cells and tissues, and to defend against this persistent attack a sophisticated system has evolved to recognize and destroy these invaders. This system comprises two components, the innate and adaptive immune systems.

The innate immune system consists of a set of genetically determined cell surface components capable of recognizing the chemical signatures of a variety of disease-causing microbes. These protective components are designed for the immediate engagement and killing of foreign organisms [15] and include several types of cells and their secreted products. On detection of an invader, white blood cells release antimicrobial agents that can kill bacteria directly or trigger other cells to ingest the organisms and degrade them. One type of white blood cell, the neutrophil, defends against infection by extruding fibers that enmesh and immobilize bacteria. These neutrophil extracellular traps have only recently been discovered and are an important component of the immune defense repertoire [16].

The other component of immunity is the adaptive immune system. Various species of scavenger cells—dendritic cells, macrophages, monocytes—patrol our bodies looking for microbial threats [17]. When these cells encounter a foreign protein, they attach this protein to a region on their surface called the major histocompatibility complex (MHC). They then travel to the lymphoid tissues, where they present the microbial protein to thymus-derived lymphocytes called T cells. T cells have CD4 molecules on their surface that can bind the MHC–microbial product complexes; this event signals for a series of changes in the T cells. They multiply, release molecules that assist in the killing of bacteria, and attach to another type of lymphocyte called the B cell. B cells produce antibodies directed against the specific invader; they also retain a memory of the encounter so that they can rapidly mount a defense if there are future exposures to this agent. CD4-bearing lymphocytes are the target of human immunodeficiency virus (HIV), and their destruction by the virus wreaks havoc on the immune system (described in later chapters).

This brief review of the immune system provides a foundation for understanding the responses people have to transfusions of blood and blood products. Blood banks attempt to closely match the red cells of the donor and recipient. However, recipients who have had multiple previous transfusions often have developed a variety of antibodies, and locating compatible donor red cells for them can be problematic. If mismatched red cells are transfused, they will be quickly destroyed and the recipient might experience chills, fever, and kidney damage. Deleterious reactions can occur not only to the red blood cells, but also to the proteins derived from the blood, such as the clotting factors used to treat hemophilia. A basic premise of the immune response is that foreign proteins are recognized as being different from the person's own proteins [18]. This response can be harmful to people with bleeding disorders, because the coagulation proteins that they receive might be identified as nonself, and the immune responses that are provoked can result in the loss of the beneficial clotting activity. In addition, the burden of these infused proteins might overwhelm cellular immune mechanisms and blunt the antibody responses to subsequent encounters with the proteins of infecting microbes. As discussed in later chapters, some hemophilia physicians had difficulty distinguishing between the immune depletion due to HIV infection and the changes in the immune system resulting from prior treatments with clotting proteins.

The next important advance in the use of blood as a therapeutic material came in the early part of the 20th century with the development of methods to prevent blood clotting [19]. These generally involved the addition of sodium citrate to the blood, but it took several years before it was recognized that the citrate solution had to be sterile to avoid inducing fever in transfusion recipients. Once blood clotting could be safely prevented, blood storage became feasible. In 1937, Fantus established the first blood bank at Cook County Hospital in Chicago [20]. It was called a bank because it was envisioned that blood would be deposited by healthy individuals and subsequently withdrawn by those in need. To maintain a positive balance, donors needed to be recruited; they could be motivated to donate either as an altruistic act or with a monetary reward. Initially, many donors were paid, but with the advent of World War II, volunteer donors came to provide the majority of the blood required [21]. After the War, the American Red Cross established regional blood centers across the United States, and they were supplemented by more than 4000 hospital and community blood banks.

During the years after World War II, the number of persons receiving blood transfusions dramatically increased due, in part, to major campaigns to recruit blood donors and the development of better methods for blood collection, storage, and administration. However, as more patients were transfused, adverse reactions became more common. Perhaps the most frequent was fever, which complicated 1–3% of all transfusions. This was due to the presence of antibodies in the patient's blood that attacked the white cells transfused with the donor blood [22]. The products released by disrupted white cells provoked chills and fever in the transfusion recipient. In addition, allergic reactions were reported in 3% of transfusions [23], possibly due to the sensitivity of the recipient to foods or drugs in the donor blood. This could occur if blood from someone who had recently eaten strawberries was given to a person allergic to this fruit. Another type of reaction, known as transfusion-related acute lung injury, occurred when a type of white cell known as a neutrophil was activated by antibodies in the donor's blood [24]. The products released by these neutrophils injured the lungs [25]. Today, these sorts of reactions have become less frequent because most donor blood is filtered to remove the white cells prior to transfusion. Currently, most severe reactions are due to human errors in typing, cross-matching, or mislabeling of blood products. Infusion of mislabeled or misdirected blood can result in the rapid destruction of the infused red cells and a fatal or near-fatal outcome for the recipient.

Blood transfusions can be harmful in other ways. For example, transfusions were shown to increase the risk of death more than twofold when given to people who have had acute heart attacks [26]. In individuals undergoing colorectal cancer surgery, transfusions during the procedure were associated with significant increases in cancer recurrence, postoperative infection, the need for surgical re-intervention, and mortality [27]. The quality of the blood transfused might also contribute to poor transfusion outcomes. Current guidelines permit storage of blood for up to 6 weeks; during storage, the capacity of the red cells to deliver oxygen declines, some of the cells break apart (hemolyze), and the risk of bacterial contamination might increase [28]. Initially, some studies showed that rates of survival after heart surgery were better in patients receiving transfused blood stored less than 2 weeks [29], but more recent trials found no greater organ dysfunction in those receiving transfusions with older versus fresh blood [30,31].

Contracting infection from contaminated blood is another major problem associated with transfusion. As previously noted, blood must be collected, stored, and infused using sterile precautions. If the donor is infected, or if the stored blood becomes contaminated by bacteria, then the organisms will be transmitted to the recipient. Transmission of hepatitis virus from donors to recipients was first reported in 1943 by Paul Beeson, who described jaundice in seven patients that appeared within 1–4 months after transfusion [32]. It was soon recognized that the risk of transfusion-transmitted hepatitis was greatest if the blood donors were drug abusers or prisoners, especially those with a history of using shared needles. These people usually donated because they were paid for their blood or, in the case of prisoners, rewarded with freedom from confinement. To decrease financial motivation for giving blood, the Red Cross and other blood collectors initiated a policy of no monetary compensation for blood donations and discontinued blood drives at prisons.

In addition to hepatitis, malaria and a variety of other infectious agents can be transmitted by transfusion [33–36a,b]. The Table lists viruses, parasites, and other microbial agents reported to have been transmitted by transfusion (coronavirus transmission is suspected but not established):

Viruses	Parasites
HIV 1 and 2	Babesiosis
Hepatitis A, B, C, D, E, and G	Chagas disease
Parvovirus B19	Malaria
Zika virus	
Coronaviruses (SARS, MERS)?	**Spirochetes**
Prions	Syphilis
TSE (mad cow disease)	

SARS, severe acute respiratory syndrome; MERS, Middle East respiratory syndrome; TSE, transmissible spongiform encephalopathy.

Because many of these microbes can reside in the tissues and blood for months to years without inducing symptoms,[1] the donors might not appear ill at the time of the blood donation. Unfortunately, there are no currently approved methods for sterilizing the blood itself, but several companies are working on various pathogen (germ) reduction techniques; some involve adding chemicals to the blood and exposing it to ultraviolet light to kill viruses and bacteria [37].

Because procedures to safely sterilize blood are still unavailable, blood banks have adopted extensive screening procedures to detect infections in donors and

---

1. For example, hepatitis viruses. Zika virus may produce no symptoms while remaining in the blood for about a week (Center for Disease Control and Prevention, quoted by C Saint Louis in *The New York Times*, February 20, 2016, p. A5.)

in the donated blood. For example, people are deferred if they have been diagnosed or treated for malaria within the previous 3 years, have an unexplained febrile illness occurring within 1 year of exposure to malaria, or even lived in a malarious area within the previous year. Potential donors now complete questionnaires and are deferred or the blood is not used for transfusion if donors are classified into any of the following categories:

Donor Deferral List 2014

- Anyone who has ever used intravenous (IV) drugs (illegal IV drugs)
- Men who have had sexual contact with other men[a]
- Anyone with a positive test result for HIV (AIDS virus)
- Men and women who have sex for money or drugs
- Anyone who has had hepatitis since age 11 years
- Anyone who has had babesiosis or Chagas disease[b]
- Anyone who has taken Tegison for psoriasis[c]
- Anyone who has risk factors for Creutzfeldt–Jakob disease (CJD) or who has a blood relative with CJD[b]
- Anyone who spent >3 months in the United Kingdom during 1980–96[d]
- Anyone who underwent transfusion in the United Kingdom or France since 1980[d]
- Anyone who has spent 5 years in Europe since 1980[d]

Notes:
[a]The FDA has proposed amending the ban on blood donation by gay and bisexual men to a 1-year deferral (2015) [38].
[b]The microbes that cause these diseases might be circulating in donor blood.
[c]Tegison might cause fetal abnormalities if transfused during pregnancy.
[d]"Mad cow disease" was reported in these countries.

Blood banks also perform extensive testing for a number of infectious agents; the currently required or recommended tests of donor blood are shown in the chart. Note that most of the testing was not initiated until 1985 (the peak year of HIV infection) or thereafter:

Agent	Year initiated
Human immunodeficiency virus	1985
Human T-cell leukemia virus	1988
Hepatitis B virus	1971
Hepatitis C virus	1990
Syphilis	1950s
West Nile virus	2003
Chagas disease agent	2007

The indications for blood transfusions include acute blood loss that is sufficiently severe to alter the function of vital organs; this might be due to surgery, injury, pregnancy, or diseases such as ulcers or cancer. Transfusions are also prescribed for individuals who are unable to make blood, and they are administered when the blood level falls to less than half the normal amount [39]. Infants born with certain heart and lung defects might also need transfusions, as do some individuals with sickle cell anemia. There are several other indications for transfusion; for example, blood is required for many cardiovascular procedures requiring the use of mechanical devices.

To collect sufficient quantities of blood to meet the various medical needs, the current completely voluntary system depends on donations by large groups from schools and businesses (blood drives), as well as donations by individuals. Because of all the mentioned reasons for donor deferral, it is estimated that only 38% of the population is eligible to give blood and only a fraction of those actually donate [40]. Extensive advertising and public service announcements are used to enlist volunteer donors (Fig. 3.1).

Currently, most physicians have adopted conservative policies for prescribing transfusions and only order blood for patients likely to fare poorly because of a low blood count [41]. The use of transfusions in the United States has declined by 30% since 2009, from 15 million units to 11 million units, and

FIGURE 3.1 Donor recruitment poster of the American Red Cross.

blood bank revenue has decreased from $5 billion to $1.5 billion [42]. There are alternatives to the use of banked blood; for example, prior to surgery anemic patients can be treated with a hormone, erythropoietin, to increase their blood counts, and during surgical procedures shed blood can be recycled back to the patient.

Blood donation is a unique form of altruistic behavior, one in which the donor does not receive the tangible rewards customarily given for other virtuous acts [43]. Most people agree to donate because they feel a sense of moral obligation and responsibility to the community [44]. Repeat donors probably also experience pleasurable internal feelings associated with the knowledge that they are contributing to restoring the health of others. In the 1970s, there was a third group of persons who regularly donated blood. These were men who had sex with men, and they constituted a sizeable percentage of repeat donors, especially on the West Coast. Their motivation to donate is unclear, but it could have been due to a need to counterbalance their negative social image with a positive contribution to society, or to show that their blood was just as good and beneficial as blood from anyone else. Those who donated blood in the late 1970s and early 1980s were almost certainly not aware that they had become infected with HIV. In the following chapters we see how blood from HIV-positive homosexual men became the fatal link between hemophilia and acquired immunodeficiency syndrome (AIDS).

KEY POINTS

- There are many references to blood in our vocabulary, religious life, and social interactions.
- Most of the knowledge about blood composition, blood groups, collection, and storage accrued during the past two centuries.
- Innate and adaptive immune systems defend against invading microorganisms, but occasionally they construe transfused blood products as foreign proteins and inactivate them.
- Technical improvements led to a remarkable increase in the number of blood transfusions after World War II.
- Recent studies suggest that the benefits of transfusion are limited, and blood conservation procedures have been widely adopted.
- It is now recognized that blood can be a vehicle for the transmission of an extensive variety of infectious agents.
- Although drug abusers and prisoners were excluded from donating blood, a sizeable percentage of the blood donors in the 1970s were men who had sex with men.
- Although there are no methods for sterilizing blood, screening of donors and donated blood for several microbial organisms is now routine, and infected units are not used for transfusion.

REFERENCES

[1] Leviticus 17:10–12.

[2] Shakespeare W. Macbeth 2.3.112.

[3] Shakespeare W. Macbeth 5.1.35; 5.1.44.

[4] Rossi EC, Simon TL, Moss GS. Transfusion in transition Rossi EC, Simon TL, Moss GS, editors. Principles of transfusion medicine. Baltimore, MD: Williams & Wilkins; 1991, p. 1.

[5] Wintrobe MM. Milestones on the path of progress Wintrobe MM, editor. Blood, pure and eloquent. New York, NY: McGraw-Hill Book Company; 1980, p. 7–11.

[6] Wintrobe MM. Milestones on the path of progress Wintrobe MM, editor. Blood, pure and eloquent. New York, NY: McGraw-Hill Book Company; 1980, p. 3–4.

[7] Legg JW. Treatise on haemophilia. London, UK: HK Lewis; 1872, p. 158.

[8] Legg JW. Treatise on haemophilia. London, UK: HK Lewis; 1872, p. 118.

[9] Wintrobe MM. Milestones on the path of progress Wintrobe MM, editor. Blood, pure and eloquent. New York, NY: McGraw-Hill Book Company; 1980, p. 17. the title page of Andral's monograph is reproduced on p. 18.

[10] London IM. Iron and heme: crucial carriers and catalysts Wintrobe MM, editor. Blood, pure and eloquent. New York, NY: McGraw-Hill Book Company; 1980, p. 173.

[11] Wintrobe MM. Milestones on the path of progress Wintrobe MM, editor. Blood, pure and eloquent. New York, NY: McGraw-Hill Book Company; 1980, p. 19–22.

[12] Rossi EC, Simon TL, Moss GS. Transfusion in transition Rossi EC, Simon TL, Moss GS, editors. Principles of transfusion medicine. Baltimore, MD: Williams & Wilkins; 1991, p. 2.

[13] Rossi EC, Simon TL, Moss GS. Transfusion in transition Rossi EC, Simon TL, Moss GS, editors. Principles of transfusion medicine. Baltimore, MD: Williams & Wilkins; 1991, p. 4.

[14] Landsteiner K. Ueber Agglutinationserscheinungen normalen menschlichen Blutes. Wien Klin Wochenschr 1901;14:1132–4.

[15] Johnston RB. An overview of the innate immune system. Waltham, MA: UpToDate, Wolters Kluwer; August 2013.

[16] Yipp BG, Kubes P. NETosis: how vital is it? Blood 2013;122:2784–94.

[17] Bonilla FA. The adaptive cellular immune response. Waltham, MA: UpToDate, Wolters Kluwer; June 2012.

[18] Tveita AA. The danger model in deciphering autoimmunity. Rheumatology 2010;49:632–9.

[19] Diamond LK. A history of blood transfusion Wintrobe MM, editor. Blood, pure and eloquent. New York, NY: McGraw-Hill Book Company; 1980, p. 677–8.

[20] Fantus B. The therapy of the Cook County Hospital. JAMA 1937;109:128–31.

[21] Diamond LK. A history of blood transfusion Wintrobe MM, editor. Blood, pure and eloquent. New York, NY: McGraw-Hill Book Company; 1980, p. 680.

[22] Lane TA, Anderson KC, Goodnough LT, Kurtz S, Moroff G, Pisciotto PT, et al. Leukocyte reduction in blood component therapy. Ann Intern Med 1992;117:151–62.

[23] Klein HG, Spahn DR, Carson JL. Red blood cell transfusion in clinical practice. Lancet 2007;370:415–26.

[24] Shaz BH, Stowell SR, Hillyer CD. Transfusion-related acute lung injury: from bedside to bench and back. Blood 2011;117:1463–71.

[25] Thomas GM, Carbo C, Curtis BR, Martinod K, Mazo IB, Schatzberg D, et al. Extracellular DNA traps are associated with the pathogenesis of TRALI in humans and mice. Blood 2012;119:6335–43.

[26] Chatterjee S, Wetterslev J, Sharma A, Lichstein E, Mukherjee D. Association of blood transfusions with increased mortality in myocardial infarction: a meta-analysis and diversity-adjusted study sequential analysis. JAMA Intern Med 2013;173:132–9.

[27]   Acheson AG, Brookes MJ, Spahn DR. Effects of allogeneic red blood cell transfusions on clinical outcomes in patients undergoing colorectal cancer surgery. Ann Surg 2012;256:235–44.

[28]   Hod EA, Brittenham GM, Billote GB, Francis RO, Ginzburg YZ, Hendrickson JE, et al. Transfusion of human volunteers with older, stored red blood cells produces extravascular hemolysis and circulating non-transferrin-bound iron. Blood 2011;118:6675–82.

[29]   Koch CG, Li L, Sessler DI, Figueroa P, Hoeltge GA, Mihaljevic T, et al. Duration of red-cell storage and complications after cardiac surgery. N Engl J Med 2008;358:1229.

[30]   Steiner ME, Ness PM, Assmann SF, Triulzi DJ, Sloan SR, Delaney M, et al. Effects of red-cell storage duration on patients undergoing cardiac surgery. N Engl J Med 2015;372:1419.

[31]   Lacroix J, Hébert PC, Fergusson DA, Tinmouth A, Cook DJ, Marshall JC, et al. Age of transfused blood in critically ill adults. N Engl J Med 2015;372(15):1410–8.

[32]   Rossi EC, Simon TL, Moss GS. Transfusion in transition Rossi EC, Simon TL, Moss GS, editors. Principles of transfusion medicine. Baltimore, MD: Williams & Wilkins; 1991, p. 8–9.

[33]   Guerrero IC, Weniger BG, Schultz MG. Transfusion malaria in the United States, 1972–1981. Ann Intern Med 1983;99:221–6.

[34]   Risseeuw-Appel IM, Kothe FC. Transfusion syphilis: a case report. Sex Transm Dis 1983;10:200–1.

[35]   Herwaldt BL, Linden JV, Bosserman E, Young C, Olkowska D, Wilson M. Transfusion-associated babesiosis in the United States: a description of cases. Ann Intern Med 2011;155:509–19.

[36a] Grant IN, Gold JW, Wittner M, Tanowitz HB, Nathan C, Mayer K, et al. Transfusion-associated acute Chagas disease acquired in the U.S.A. Ann Intern Med 1989;111:849–51.

[36b] Nickerson P, et al. Transfusion-associated *Trypanosoma cruzi* infection. Ann Intern Med 1989;111:851–3.

[37]   Snyder EL, Stramer SL, Benjamin RJ. The safety of the blood supply-time to raise the bar. N Engl J Med 2015;372:1882–4.

[38]   Hamburg MA. FDA Commissioner Margaret A Hamburg's statement on FDA's blood donor deferral policy for men who have sex with men. U.S. Food and Drug Administration Press Announcement; December 23, 2014.

[39]   Carson JL, Grossman BJ, Kleinman S, et al. Red blood cell transfusion: a clinical practice guideline from the AABB. Ann Intern Med 2012;157:49–58.

[40]   Donor recruiting information from the American Red Cross; September 2014.

[41]   Holst LB, Petersen MW, Haase N, Perner A, Wetterslev J. Restrictive versus liberal transfusion strategy for red blood cell transfusion: systematic review of randomized trials with meta-analysis and trial sequential analysis. BMJ 2015;350:h1354.

[42]   Wald ML. Blood industry hurt by surplus. NY Times 2014;August 23.

[43]   Titmuss RM. The gift relationship: from human blood to social policy. New York, NY: Vintage Books; 1971.

[44]   Andre C, Velasquez M. Giving blood: the development of generosity. Markkula Center for Applied Ethics, Santa Clara University, CA; 2014.

Chapter 4

The Human Immunodeficiency Virus

The acquired immunodeficiency syndrome (AIDS) was first reported in the United States in the early 1980s, but the causative agent, human immuno-deficiency virus (HIV), had actually infected people several decades earlier. Scientists have examined archival materials from the earliest known patients with AIDS as well as tissues and fluids from African primates for evidence of HIV infection. Because the virus has undergone a series of genetic changes (mutations) over many decades, it is possible to determine when the earliest ancestor of the virus appeared and when each subspecies became prevalent; this is called phylogenetic analysis. Using this method, a closely related virus (simian immunodeficiency virus) first appeared in the Cameroons in 1910, and it probably spread from chimpanzees to humans through exposure to infected ape blood and/or body fluids [1]. In approximately 1920, a person presumably infected by the virus traveled down the Sangha River to Kinshasa (formerly known as Leopoldville), the capital of the Democratic Republic of the Congo (previously called Zaire) [2]. That Kinshasa was the geographical center of the infection is supported by studies identifying the genetic signature of HIV in tissue removed in 1960 from a person living in that city [3]. There is evidence that the virus was also transported along Zairian railway lines and waterways, eventually reaching most of the major cities in that country [4]. One of the earliest cases of AIDS, reported in 1977, described a Zairian woman who developed fever, thrush, weight loss, and enlarged glands associated with repeated infections, including meningitis [5]. Severe immunodeficiency was confirmed by laboratory tests and she died in early 1978. By 1982, at least a dozen patients had been admitted to hospitals in Zaire with illnesses suggestive of AIDS.

In the early 1970s Belgians were expelled from Zaire, and they might have brought the virus with them on their return to Europe (Fig. 4.1). One of these individuals could have transmitted it to a homosexual German violinist, who was discovered to be infected in late 1976; this man had traveled exten-sively throughout the continent but had never visited Africa [6]. The virus that reached the Americas took a different route. Early in the 1970s, Zairian lead-ers invited trained professionals from Haiti to assist the developing nation with

Linked by Blood: Hemophilia and AIDS. DOI: http://dx.doi.org/10.1016/B978-0-12-805302-7.00004-5

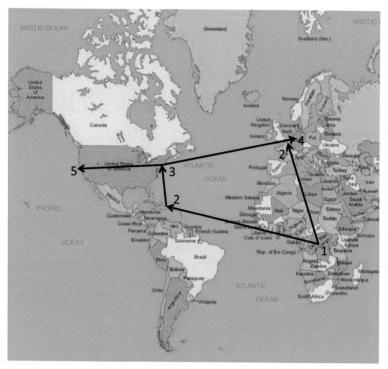

FIGURE 4.1 AIDS flight path. (1) Origin in Zaire (Democratic Republic of Congo): 1960; (2) Spread to Haiti and Belgium: 1970s; (3) Spread to New York: 1976; (4) Spread to Europe: 1980s; (5) Spread to San Francisco: 1980s.

administrative and other tasks. While working in Zaire, the visiting Haitians probably acquired HIV from infected locals, and then transmitted it to other Haitians on their return to Hispaniola [7]. Subsequently, the virus spread to American and European tourists visiting the island nation. Another potential point of entry of the virus into the United States was the bicentennial celebration of 1976, when tall sailing ships and their crews from 55 nations came to New York [8]. Infected sailors might have introduced the virus to their sexual partners. Some researchers have estimated that the virus could have been circulating in the US population beginning as early as 1966 [9], but its disease-causing effects were not recognized until the first cases of AIDS were reported in 1978.

As physicians began seeing patients with unusual infections, they reported them to the Centers for Disease Control (CDC). By April 1982, the CDC counted 300 infected persons and 119 deaths. Because mostly gay or bisexual men were affected, it was thought that their unique lifestyle led to immune defects. However, as the number of patients with AIDS progressively increased, it was recognized that the syndrome was transmissible; women could acquire AIDS from their male partners, hemophiliacs and others from blood products, and infants could acquire AIDS from their mothers [10]. The cause of AIDS was unknown, but most researchers suspected a virus was involved. Similarities to the epidemiology of hepatitis B virus infection led to a theory that the AIDS agent was carried by the hepatitis virus [11]. However, this was unlikely because a study involving hemophiliacs showed no association between liver disease caused by hepatitis virus and immunologic abnormalities associated with AIDS [12]. Other suspects were herpes simplex virus, Epstein–Barr virus, cytomegalovirus, and even the stimulant amyl nitrite, which was used by drug abusers. Researchers searched for exotic viruses, parasites, and other pathogens in tissues from patients with AIDS. An unusual inclusion body was reported in the cytoplasm of lymphoid cells obtained from homosexual men [13], and other workers noted tubule-reticular structures, rings, and test tube–shaped forms [14], but it was unclear if these were normal cellular components [15], invading organisms, membranes of activated cells, or even artifacts of sample preparations.

A disease similar to AIDS was identified in a colony of monkeys; four separate outbreaks were described. The animals developed enlarged lymph glands and spleens, had diarrhea, fever, and anemia, and experienced multiple infections similar to those observed in human patients with AIDS [16]. The disease could be transmitted to healthy animals by injecting tissue fluids from sick animals [17,18] or by inoculation of filtered blood or plasma, suggesting that the causative agent was most likely a virus [19].

A series of articles in the May 20, 1983 issue of *Science* magazine convincingly linked human AIDS to a virus. The laboratories of Gallo and Essex in the United States and of Montagnier in France isolated a human retrovirus from patients with AIDS. A retrovirus produces DNA from its RNA genome, the reverse of the usual pattern, and inserts this DNA into the DNA of the host. The researchers observed that infected individuals were more likely to have antibodies to this retrovirus than noninfected persons [20–22]. However, they could not exclude the possibility that the virus was just another microbe that took advantage of the diminished immune system of these people. Nevertheless, there were several lines of evidence supporting the theory that a retrovirus was responsible for AIDS. It was prevalent in the Caribbean, including Haiti, where AIDS was reported; it was known to infect T-cells, which were decreased in patients with AIDS; and it could be transmitted by close personal contact, consistent with the sexual transmission of the new disease. Gallo thought that this retrovirus might be the human T-cell leukemia virus (HTLV-1) that he had discovered several years previously. However, there were several critical observations that made this an unlikely possibility. HTLV-1 was detected in only 2 of 33 AIDS patients, and it could not be recovered in subsequent blood samples from these same patients. HTLV-1 was common in southern Japan, but AIDS was absent in this region. In addition, patients known to be infected by HTLV-1 developed T-cell leukemia rather than AIDS. In fact, when lymphoid cells were cultured in the presence of the new virus, they were killed; in contrast, HTLV-1 actually caused the cells to multiply. For these and other reasons, many doubted that HTLV-1 was the cause of AIDS [23].

The concerns about the authenticity of HTLV-1 as the cause of AIDS led the Gallo group to propose that the virus they isolated was not HTLV-1, but a related virus, which they termed HTLV-III [24]. They developed a serologic test that was very specific for this virus; all individuals with clear evidence of AIDS tested positive, whereas healthy persons were negative [25]. The Montagnier group did not believe that the virus they had isolated from the lymph node of an AIDS patient was HTLV, and they named it the "lymphadenopathy associated virus" or LAV [26]; it appeared to have the same properties as HTLV-III [27]. A third virus, called AIDS-associated retrovirus (ARV-34), was isolated by workers in San Francisco from homosexual males with AIDS and a man with hemophilia [28]. Antibodies to this virus were detected in the serum of 28 hemophiliacs treated with factor VIII concentrate. Its properties were similar to HTLV-III and LAV, and eventually all these viruses were included in the sobriquet HIV (Figs. 4.2 and 4.3) [29].

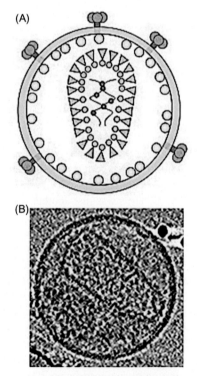

FIGURE 4.2   (A): Schematic model showing the organization of the mature HIV-1 virion. (B): Central section from a tomographic reconstruction of a mature HIV-1 virion. *Reproduced with permission from Sundquist WI, Krausslich HG. HIV-1 assembly, budding, and maturation. Cold Spring Harb Perspect Med July/August 2012;2(7):a006924.*

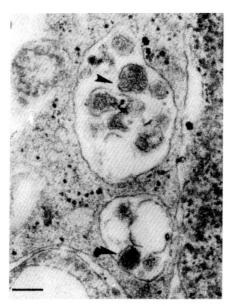

FIGURE 4.3   Viruses (*arrows*) in the cells of a 55-year-old man with hemophilia. *Reproduced with permission from Palmer EL, et al. Human T-cell leukemia virus in lymphocytes of two hemophiliacs with the acquired immunodeficiency syndrome. Ann Intern Med 1984;101:293–7.*

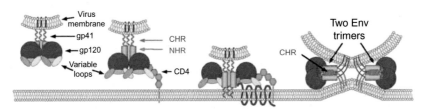

FIGURE 4.4   A schematic of the HIV surface protein, gp120, locating the receptor on a CD4 lymphocyte, followed by the insertion of the viral gp41 protein into the CD4 cell membrane. In the last image, the viral membrane has fused with the lymphocyte membrane. *Reproduced with permission from Klasse PJ. The molecular basis of HIV entry. Cell Microbiol 2012;14:1183.*

HIV infects host cells in a manner analogous to the way mosquitoes suck blood from their prey. The virus has a surface protein designated gp120, which seeks host cells bearing a membrane marker called CD4. Once the gp120 has attached to the CD4, a second viral protein, gp41, penetrates the cell membrane, much like the proboscis of the mosquito penetrates the skin. But instead of drawing blood from the wound, the virus pours its genetic material into the cell (Fig. 4.4). The viral RNA then commandeers the cell machinery and forces it to make new viral particles. Eventually the cell ruptures, releasing thousands of new viruses ready to attack the host's cells. Even those cells able to resist viral proliferation succumb as a consequence of the acute inflammatory reaction provoked by the presence of the virus [30]. As a consequence of the relentless attack by HIV, CD4-bearing T-cells, which normally fight infections, become depleted and resistance to infection is impaired. During the 1980s, laboratories measured the number of CD4 T-cells and the ratio of these cells to CD8 T-cells. Normally, CD4 T-cells are more numerous than CD8 T-cells and the ratio is greater than 1.0, but it is almost always less than 1.0 in HIV-infected persons. A low CD4:CD8 ratio is also occasionally observed in other diseases; therefore, it is not specific for HIV infection, but it does indicate that the immune system has been compromised. Currently, HIV infection is diagnosed by testing for the presence of antibodies to the virus, and the response to treatment is monitored by measuring the number of viral particles in the blood.

HIV transmission usually occurs through unprotected intercourse. An HIV-infected man has virus in his semen - the virus might be either, in the fluid itself, in the sperm, or in the white blood cells that are often present in the fluid [31]. During heterosexual intercourse, the virus penetrates the tissues of a man's partner through the small abrasions that occur in the vaginal lining; the virus remains in the local area for approximately 1 week before it begins to spread through the rest of her body. If a woman infected by HIV has intercourse with an uninfected man, then the virus in her vaginal secretions might find refuge under his foreskin if he is uncircumcised; studies have shown that circumcision significantly decreases the incidence of HIV [32]. Another route for viral transmission is provided when men have sex with men; during anal intercourse, the delicate rectal lining is easily abraded, allowing penetration by the virus. It has been estimated that HIV infection is 50-times more likely with anal intercourse than with vaginal intercourse [21]. The coexistence of other sexually transmitted diseases greatly enhances the risk of becoming infected by HIV.

HIV can also be transmitted from an infected pregnant woman to her fetus, either in utero or when there is an exchange of bloody fluids during childbirth. To prevent the transmission of HIV from mother to child, pregnant women are routinely tested for HIV infection, and those with positive test results are given antiretroviral therapy. This strategy has greatly reduced the frequency of mother-to-child transfer of HIV. Another group of individuals susceptible to HIV infection are intravenous drug abusers, whose dirty needles and syringes are readily contaminated by the blood-borne virus. Transmission of the virus among those in the subculture of prostitutes, pimps, and drug abusers accounts for much of the heterosexual spread of the disease. In one study of seven drug-using men with AIDS, investigators found that six of their female sexual partners had either overt AIDS or immunologic abnormalities typical of HIV-induced immunodeficiency [33].

Health professionals who treat HIV-infected patients are at risk because of inadvertent needle punctures and exposures to infected blood during operations. One of the earliest recognized cases of AIDS occurred in a Danish doctor working in Zaire, Africa, from 1972 to 1977 [34]. This surgeon operated under primitive conditions and her bare hands were often immersed in patients' blood. She could recall seeing at least one patient with Kaposi's sarcoma, a tumor chiefly associated with HIV infection (Fig. 4.5). In 1976, the doctor developed diarrhea, fatigue, weight loss, and enlarged glands; her symptoms progressed, and later the next year she died of pneumonia. A postmortem examination showed *Pneumocystis jirovecii* in her lungs, an opportunistic infection that commonly occurs in people with AIDS. Exposure to infected blood might also have been responsible for the AIDS-related death of an anthropologist who had taken part in a ritual interchange of blood with East African tribesmen—a "blood brotherhood" ceremony [35]. It is important to note, however, that HIV is incapable of surviving outside of bodily fluids, and it is not transmitted by kissing, touching, or contact with tears or saliva.

Infection with HIV occurs in two phases. The initial exposure is associated with fever and sweating, muscle and joint pains, sore throat, loss of appetite, headaches, and occasionally a rash [36]. These symptoms last for 3–14 days, and they do not distinguish HIV infection from many other acute viral illnesses. The second phase begins with gradual swelling of the lymph glands. The onset of AIDS is subtle and mainly consists of skin rashes resembling dandruff or acne, whitish plaques or cheese-like material on the tongue and lining of the mouth and throat, diarrhea, fever, and weight loss. In a young person, these signs and symptoms are often ignored because they usually do not directly interfere with school or work. However, after many months or years of living with HIV, the individual will develop a serious infection and seek medical attention. The infections are due to "opportunistic" organisms; these are bacteria, fungi, and parasites that usually do not cause illness in persons with adequate numbers of CD4-bearing immune cells. One such fungus, *Pneumocystis jirovecii*, is associated with a distinctive form of pneumonia that has become the hallmark of AIDS. Another disease, Kaposi's sarcoma, consists of multiple skin tumors (Fig. 4.5). These conditions are rarely encountered in persons without HIV infection. Other cancers are also more common in HIV-infected people, especially cancers of the lymph glands called lymphomas [37,38].

Without treatment, most HIV-infected persons die of either infection or cancer. Of great interest are the 0.5% whose disease advances slowly or not at all; they are termed "nonprogressors" or "HIV controllers," and they appear to remain well for years if not indefinitely [39]. Proliferation of the virus is minimal or undetectable in these individuals. Extensive studies have shown that some nonprogressors have genetic mutations in cell-surface proteins, such as those concerned with self-identification (human leukocyte antigen) or immune modulation (chemokine receptor type 5), that block binding of the virus to CD4 cells. Others have highly potent CD8 T-cells and dendritic cells that restrict viral replication. How these infected persons are able to maintain levels of CD4 T-cells and immune activation similar to those of noninfected individuals is currently unknown but might provide important clues for protecting others from the

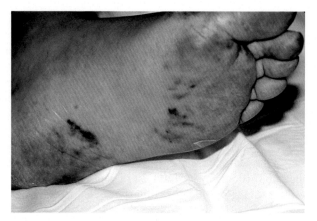

FIGURE 4.5 Kaposi's sarcoma involving the foot. *Photograph courtesy of the author.*

ravages of this disease. An important feature of HIV infection is the long delay between the initial infection and the appearance of AIDS. Epidemiologists at the CDC observed a cluster of cases in homosexual men in early 1982, and they suspected that these men had probably been exposed to the virus in 1979 [40]. However, in Dec. 1983, they were able to assess the length of the incubation period more accurately by measuring the time from transfusion of contaminated blood to the development of AIDS; in 21 persons, the average incubation period between the time of exposure to the infected blood and the appearance of AIDS was a lengthy 5.5 years. Knowledge that the disease had a long incubation period was of crucial importance; it dispelled the notion that because full-blown AIDS had been diagnosed in only a few patients, there was no epidemic. Rather, it indicated that there could be a large number of people infected by the virus, but they might not exhibit signs and symptoms of the disease for several years. In fact, there were warnings that the known cases of AIDS were only "the tip of the iceberg" of what was to come.

By 1983, 2 years after the first descriptions of *Pneumocystis* pneumonia in young homosexual men, more than 1000 definite cases of serious opportunistic infections with or without Kaposi's sarcoma had been reported from 33 states and 13 foreign countries [41]. Although most of those with the disease were homosexual men or intravenous drug abusers, others were immigrants from Haiti who were neither homosexuals nor drug addicts [42]. In fact, Haiti appeared to be a hot bed of AIDS; as many as 29 persons with the disease were described in a report from Port-au-Prince [43]. Hemophiliacs treated with blood products constituted another group affected; of greatest concern was that even hemophiliacs who appeared well were found to have depletion of CD4 cells, often to the same extent as observed in infected homosexual men [44]. Because there was no definitive test for HIV at this time (1983), the cause of the immunodeficiency was unclear, but some thought the pattern of transmission was reminiscent of infection with hepatitis B virus [45]. All disease transmissions appeared to occur by sexual contact, contaminated needles, or blood components; there were no cases of AIDS among health care or laboratory workers exposed to infected patients unless their skin was pierced by contaminated needles. Therefore, it was recommended that the precautions then in use for hepatitis B, which was also mainly transmitted by contaminated blood or needles, should be extended to AIDS [46]. These included warning labels on patient specimens, use of gown and gloves for handling blood and other secretions, protective eyewear if splattering of blood was anticipated, and thorough handwashing before and after contact with infected persons [47]. Patients did not have to be isolated unless potential roommates were vulnerable to infection. Most importantly, it was recommended that health personnel, unless pregnant or ill, should not be excused from delivering care to AIDS patients, but should not provide mouth-to-mouth resuscitation. Routine screening of personnel or patients for HIV was not recommended, but it was suggested that counseling and HIV antibody testing should be offered to pregnant women at high risk [48]. Emphasis was placed on the development of intensive and continuing educational programs involving the epidemiology of HIV infection and appropriate infection control practices.

In May 1983, a Mini Brief on AIDS, prepared by Judith A. Johnson of the Library of Congress Congressional Research Service, reported that the death toll from AIDS stood at 520, and that the disease appeared to be spreading from homosexuals to hemophiliacs, children of AIDS victims, and blood transfusion recipients [49]. Another report from the CDC observed that the costs for the first 1000 AIDS cases were estimated at $60 million in hospital expenses alone. On April 26, 1983, the Social Security Administration issued emergency instructions ordering disability benefits for all applicants with AIDS-related infections who were unable to work but had paid enough payroll taxes to qualify.

Because the cause of AIDS was unknown and all treatments were experimental, federal health agencies were under great pressure to control the epidemic. But in 1981, Ronald Reagan was elected on a platform that promised to cut the federal budget, and the Office of Management and Budget proposed to decrease funding for the CDC from the $327 million recommended in the Carter budget to only $161 million [50]. Despite the decreased budget, the CDC increased its research funding from $200,000 in FY1981 to $2.050 million in FY1982, and expected to spend an estimated $4.6 million in FY1983 and at least $4.3 million in FY1984 [38]. However, the Reagan Administration's FY1984 research budget included only $2 million for the CDC. Underfunding for AIDS research characterized the early years of the epidemic and resulted in hiring freezes, fewer research grants, and the exodus of government scientists to industry. Of note, federal spending, recorded as expenditure per death, was four-times greater for Legionnaire's disease and toxic shock syndrome than for AIDS.

The *Morbidity and Mortality Weekly Report* of June 1984 displayed a chart showing the number of AIDS cases from 1981 to 1984 [51]. The rapid increase during this period is evident; the total number of cases reached 4918, and 45% were known to have died. More than 76% of patients diagnosed prior to July 1982 were dead. The total included 42 hemophiliacs and 52 adults whose only risk factor was transfusion of blood or blood components.

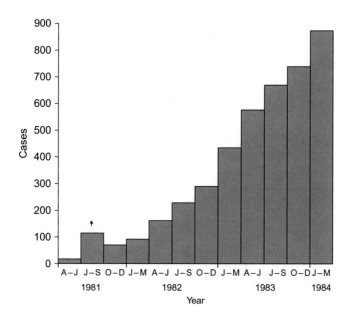

The growing concern about the AIDS epidemic eventually led to a major increase in funding, which increased from $5.5 million in 1982 to $42,489,000 in 1984 [52]. This financial support was critical for future discoveries about the disease. It led to the identification of the putative causative agent by several groups as previously described, and the development of diagnostic laboratory tests. These showed that the percentage of homosexual men testing positive for the virus progressively increased with increasing numbers of male sexual partners (fewer than 10, 25%; 11–50, 35%; more than 50, 63%) [53], clearly revealing the risks of sexual promiscuity. However, casual contact with infected persons was not a risk factor; all personnel working with specimens from hemophiliacs tested negative [54].

Progress in the development of effective drugs for AIDS was slow, but by October 1985, several compounds were under active investigation. These included antiviral agents and drugs that stimulated the immune system such as interferons [55] and interleukins, but these medications had considerable toxicity and were unsuitable for widespread use [56]. There was considerable interest in suramin, a drug used to treat certain parasitic diseases that was reported to limit HTLV-III infectivity and T-cell death [57]. However, a clinical trial of suramin involving 10 patients with AIDS showed no significant clinical or immunological improvement; in addition, the drug provoked fever, rashes, and liver inflammation [58], leading to its abandonment.

In late 1985, the Burroughs Wellcome Company began testing azidothymidine,[1] a compound that prevents viral multiplication, and soon reported that the drug significantly decreased HIV levels [59]. The drug needed to be taken every 4 hours around the clock and caused a decrease in blood counts, but it was generally well-tolerated and was approved by the Food and Drug Administration (FDA) in 1986. Later studies showed that it significantly reduced the number of patients developing AIDS [60]. Another therapeutic advance was the recognition that relatively low doses of an antibiotic called trimethoprim–sulfamethoxazole, when taken daily, could prevent HIV-infected persons from developing *Pneumocystis* pneumonia [61].

In the decades following the advent of the HIV epidemic, great strides were made in the diagnosis and treatment of this infection. However, millions of people had become infected with the virus; in 1999, almost one in four pregnant women in South Africa was HIV-positive [62]. The United Nations recently estimated that in 2012, there were 2.3 million new HIV infections in children and adults and 1.6 million AIDS-related deaths [63]. To combat the epidemic, the US President's Emergency Plan for AIDS Relief (PEPFAR) was implemented in 2003. This program has provided an estimated 5 million patients in the developing world with antiretroviral drugs and protected nearly 1 million infants from maternally transmitted HIV [64].

Persons with AIDS and their partners pioneered a direct form of advocacy, demanding a "place at the table" in the planning and development of treatment for the disease. They offered proposals to accelerate research, modify clinical trials, and improve regulatory processes, and they played a major role in fundraising. In addition, they have campaigned to reduce the cost of essential medicines, both in the United States and abroad. This has resulted in rulings permitting the production of generic equivalents of patented AIDS medications in countries such as India, making these drugs much more widely available. The advent of the AIDS epidemic has led to a new paradigm, Global Health, for confronting diseases that cross national borders, and to the introduction of research and educational programs targeted to these disorders. A pioneer in this effort was Joep Lange, who co-founded several organizations to provide global access to information and medications for AIDS [65]. As Allan Brandt has noted, "AIDS has reshaped conventional wisdoms in public health, research practice, cultural attitudes, and social behaviors" [66].

However, HIV continues to infect as many as 50,000 persons per year in the United States, and it is estimated that more than 34 million people worldwide are currently infected. To combat this ongoing epidemic, the CDC has recommended that healthy persons at risk for contracting HIV infection should take daily doses of

---

1. Azidothymidine is the chemical name, zidovudine is the generic name, Retrovir is the brand name, and AZT is the abbreviated name.

an antiretroviral pill in conjunction with condom use [67]. The doses recommended for this prophylaxis are smaller than those used to treat the infection, and the cost is covered by insurance. This is a most welcome development for the spouses of HIV-infected men with hemophilia. Another advance has been the development of a once-daily combination pill; this medication has been shown to eliminate detectable blood-borne virus in 88% of patients and has minimal side effects. However, the problem of virus persisting in immune cells has yet to be overcome [68,69].

There are many lessons to be learned from the AIDS epidemic. First, it showed the tremendous reach of infectious diseases in our highly connected world; international travel has greatly facilitated the spread of microbes [70]. Second, an organism (HIV) was encountered with properties unlike any of those previously known; the unusual behavior of the virus confused experts and delayed recognition that it was the agent responsible for AIDS. It differed from familiar viruses such as those causing measles and mumps in its ability to remain dormant for prolonged periods and then emerge and deplete the cells that guard against other infectious organisms. So far, this behavior has thwarted attempts at curative therapy and the production of a protective vaccine. It has also been learned that the discovery of a safe and effective drug is not sufficient to curb an epidemic; people must also be educated about the signs and symptoms of the disease so that they will seek medical assistance, and the drugs that are prescribed must be accessible to all those infected. Although it is important to compensate pharmaceutical companies for the expenses of drug development and manufacture, drug company profits must not stand in the way of restoring health to infected persons and preventing the spread of disease. Finally, we have learned that the research enterprise must be fully supported, because new discoveries will be needed to deal with emerging pathogens.

KEY POINTS

- A simian virus was discovered in the Cameroons in 1910, and it subsequently spread from chimpanzees to humans in the 1940s, probably through exposure to infected ape blood. Human immunodeficiency virus (HIV) was identified in human tissue from the Belgian Congo in 1960.
- Haitian administrators invited to the Belgian Congo after the 1960s rebellion became infected and, on their return home, transmitted the infection to other Haitians. The virus spread to American tourists in the 1970s. In addition, the US bicentennial of 1976 attracted sailors to New York who were infected by the virus and likely disseminated it to the gay community.
- The initial infection presents as a typical flu-like illness, and it is followed by a quiescent period averaging 5.5 years before the appearance of acquired immunodeficiency syndrome (AIDS).
- During the quiescent period, the virus multiplies and helper lymphocytes (CD4+ cells) are depleted, so that the ratio of CD4 to CD8 lymphocytes declines from 2 to <0.5, enabling the development of a variety of

opportunistic infections. Chief among these are *Pneumocystis jirovecii* and Kaposi's sarcoma, although the latter is infrequent in hemophiliacs. Death is usually due to organisms such as *P. jirovecii*, *Mycoplasma intracellulare*, and *Cryptococci*, as well as lymphoma and other cancers.

- An assay for HIV antibody was not available until late 1984; therefore, prior to that time, the actual extent of the epidemic was unknown. Currently, the amount of virus in the blood can be accurately measured.
- The infection is treated with antiretroviral agents. The first drug to be approved was azidothymidine (AZT) in 1986, and now there are several more effective agents with fewer side effects.

REFERENCES

[1] Sharp PM, Hahn BH. Origins of HIV and the AIDS pandemic. Cold Spring Harb Perspect Med 2011;1:a006841.

[2] Cohen J. Early AIDS virus may have ridden Africa's rails. Science 2014;346:21–2.

[3] Worobey M, Gemmel M, Teuwen DE, et al. Direct evidence of extensive diversity of HIV-1 in Kinshasa by 1960. Nature 2008;455:661–4.

[4] Faria NR, Rambaut A, Suchard MA, et al. The early spread and epidemic ignition of HIV-1 in human populations. Science 2014;346:56–61.

[5] Vandepitte J, Verwilghen R, Zachee P. AIDS and cryptococcosis (Zaire, 1977). Lancet 1983;i:925.

[6] Sterry W, Marmor M, Konrads A, Steigleder GK. Kaposi's sarcoma, aplastic pancytopenia, and multiple infections in a homosexual (Cologne, 1976). Lancet 1983;i:924–5.

[7] Gilbert MT, Rambaut A, Wlasiuk G, Spira TH, Pitchenik AE, Worobey M. The emergence of HIV/AIDS in the Americas and beyond. Proc Natl Acad Sci USA 2007;104:18566–70.

[8] Shilts R. And the band played on. New York, NY: St Martin's Press; 1987, p. 630.

[9] Li W-H, Tanimura M, Sharp PM. Rates and dates of divergence between AIDS virus nucleotide sequences. Mol Biol Evol 1988;5:313–30.

[10] Anonymous. Acquired immunodeficiency syndrome. Lancet 1983;i:162–3.

[11a] McDonald MI, Hamilton JD, Durack DT. Hepatitis B surface antigen could harbor the infective agent of AIDS. Lancet 1983;ii:882–4.

[11b] Ravenholt RT. Role of hepatitis B virus in acquired immunodeficiency syndrome. Lancet 1983;ii:885.

[12] Froebel KS, Lowe GDO, Madhok R, Forbes CD. AIDS and hepatitis B. Lancet 1984;i:632.

[13a] Ewing Jr EP, Spira TJ, Chandler FW, Callaway CS, Brynes RK, Chan WC. Unusual cytoplasmic body in lymphoid cells of homosexual men with unexplained lymphadenopathy. N Engl J Med 1983;308:819–22.

[13b] Zucker-Franklin D. Looking for the cause of AIDS. N Engl J Med 1983;308:837–8.

[14a] Sidhu GS, Stahl RE, El-Sadr W, Zolla-Pazner S. Ultrastructural markers of AIDS. Lancet 1983;i:990–1.

[14b] Orenstein JM, Ewing Jr EP, Spira TJ, Chandler FW, Callaway CS, Brynes RK, et al. Ultrastructural markers of AIDS. Lancet 1983;ii:284–5.

[15] Gardiner T, Kirk J, Dermott E. "Virus-like particles" in lymphocytes in AIDS are normal organelles, not viruses. Lancet 1983;ii:963–4.

[16] Henrickson RV, Osborn KG, Madden DL, et al. Epidemic of acquired immunodeficiency in rhesus monkeys. Lancet 1983;i:388–90.

[17]   Letvin NL, King NW, Daniel MD, Aldrich WR, Blake BJ, Hunt RD. Experimental transmission of macaque AIDS by means of inoculation of macaque lymphoma tissue. Lancet 1983;ii:599–602.

[18]   London WT, Sever JL, Madden DL, et al. Experimental transmission of simian acquired immunodeficiency syndrome (SAIDS) and Kaposi-like skin lesions. Lancet 1983;ii:869–73.

[19]   Gravell M, London WT, Houff SA, et al. Transmission of simian acquired immunodeficiency syndrome (SAIDS) with blood or filtered plasma. Science 1984;223:74–6.

[20]   Gallo R, Sarin PS, Gelmann EP, et al. Isolation of human T-cell leukemia virus in acquired immune deficiency syndrome (AIDS). Science 1983;220:806–9; 865–7.

[21]   Essex M, McLane MF, Lee TH, et al. Antibodies to cell membrane antigens associated with human T-cell leukemia virus in patients with AIDS. Science 1983;220:859–62.

[22]   Barre-Sinoussi F, Chermann JC, Rey F, et al. Isolation of a T-lymphotropic retrovirus from a patient at risk for acquired immune deficiency syndrome (AIDS). Science 1983;220:868–71.

[23]   Black PH, Levy EM. The human T-cell leukemia virus and AIDS. N Engl J Med 1983;309:856.

[24a] Popovic M, Sarngadharan MG, Read E, Gallo RC. Detection, isolation, and continuous production of cytopathic retroviruses (HTLV-III) from patients with AIDS and pre-AIDS. Science 1984;224:497–500.

[24b] Gallo RC, Salahuddin SZ, Popovic M, et al. Frequent detection and isolation of cytopathic retroviruses (HTLV-III) from patients with AIDS and at risk for AIDS. Science 1984;224:500–3.

[25]   Safai B, Groopman JE, Popovic M, et al. Seroepidemiological studies of human T-lymphotropic retrovirus type III in acquired immunodeficiency syndrome. Lancet 1984;i:1438–40.

[26]   Vilmer E, Barre-Sinoussi F, Rouzioux C, et al. Isolation of new lymphotropic retrovirus from two siblings with haemophilia B, one with AIDS. Lancet 1984;i:753–7.

[27]   Feorino PM, Kalyanaraman VS, Haverkos HW, et al. Lymphadenopathy associated virus infection of a blood donor-recipient pair with acquired immunodeficiency syndrome. Science 1984;225:69–72.

[28a] Levy JA, Hoffman AD, Kramer SM, Landis JA, Shimabukuro JM, Oshiro LS. Isolation of lymphocytopathic retroviruses from San Francisco patients with AIDS. Science 1984;225:840–2.

[28b] Koerper MA, Kaminsky LS, Levy JA. Differential prevalence of antibody to AIDS-associated retrovirus in haemophiliacs treated with factor VIII concentrate versus cryoprecipitate: recovery of infectious virus. Lancet 1985;i:275.

[29]   Coffin J, Haase A, Levy JA, et al. Human immunodeficiency viruses [letter]. Science 1986;232:697.

[30]   Gaiha GD, Brass AL. The fiery side of HIV-induced T-cell death. Science 2014;343:383–4.

[31]   Sabatte J, Lenicov FR, Cabrini M, et al. The role of semen in sexual transmission of HIV: beyond a carrier for virus particles. Microbes Infect 2011;13:977–82.

[32]   Reed JB, Njeuhmeli E, Thomas AG, et al. Voluntary medical male circumcision: an HIV prevention priority for PEPFAR. J Acquir Immune Defic Syndr 2012;60:S88–95.

[33]   Harris C, Small CB, Klein RS, et al. Immunodeficiency in female sexual partners of men with the acquired immunodeficiency syndrome. N Engl J Med 1983;308:1181–4.

[34]   Bygbjerg IC. AIDS in a Danish surgeon (Zaire, 1976). Lancet 1983;i:925.

[35]   Morfeldt-Manson L, Lingquist L. Blood brotherhood: a risk factor for AIDS? Lancet 1984;ii:1346.

[36]   Cooper DA, Maclean P, Finlayson R, et al. Acute AIDS retrovirus infection. Lancet 1985;i:537–40.

[37]   Ragni MV, Lewis JH, Bontempo FA, Spero JA. Lymphoma presenting as a traumatic hematoma in an HTLV-III antibody-positive hemophiliac. N Engl J Med 1985;313:640.

[38] Kaplan MH, Susin M, Pahwa SG, et al. Neoplastic complications of HTLV-III infection. Am J Med 1987;82:389–96.

[39] Walker BD, Yu XG. Unravelling the mechanisms of durable control of HIV-1. Nat Rev Immunol 2013;13:487–98.

[40] The events and key figures that established the HIV incubation period are described by Shilts R. And the band played on. New York, NY: St Martin's Press; 1987, pp. 132, 402.

[41] Curran JW, Evatt BL, Lawrence DN. Acquired immune deficiency syndrome: the past as prologue. Ann Intern Med 1983;98:401–3.

[42] Vieira J, Frank E, Spira TJ, Landesman SH. Acquired immune deficiency in Haitians. N Engl J Med 1983;308:125–9.

[43] Malebranche R, Guerin JM, Laroche AC, et al. Acquired immunodeficiency syndrome with severe gastrointestinal manifestations in Haiti. Lancet 1983;ii:873–8.

[44] Goldsmith JC, Moseley PL, Monick M, Brady M, Hunninghake GW. T-lymphocyte subpopulation abnormalities in apparently healthy patients with hemophilia. Ann Intern Med 1983;98:294–6.

[45] Curran JW. AIDS—two years later. N Engl J Med 1983;309:609–11.

[46] Conte Jr JE, Hadley WK, Sande M. Infection-control guidelines for patients with the acquired immunodeficiency syndrome (AIDS). N Engl J Med 1983;309:740–4.

[47] Anonymous. Summary: recommendations for preventing transmission of infection with human T-lymphotropic virus type III/lymphadenopathy-associated virus in the workplace. MMWR Morb Mortal Wkly Rep 1985;34:681–94.

[48] Conte Jr. JE. Infection with human immunodeficiency virus in the hospital. Ann Intern Med 1986;105:730–6.

[49] Johnson JA. AIDS: Acquired immunodeficiency syndrome: Mini Brief Number MB83211. The Library of Congress, Congressional Research Service, Major Issues System; Date updated May 31, 1983.

[50] Shilts R. And the band played on. New York, NY: St Martin's Press; 1987, p. 55.

[51] Anonymous. Update: acquired immunodeficiency syndrome (AIDS)-United States. MMWR Morb Mortal Wkly Rep 1984;33:337.

[52] National Hemophilia Foundation. Increased 1984 funding for National Institutes of Health and CDC AIDS research. Hemophilia Information Exchange; October 21, 1983.

[53] Anderson RE, Levy JA. Prevalence of antibodies to AIDS-associated retrovirus in single men in San Francisco. Lancet 1985;i:217.

[54] Jones P, Hamilton P. HTLV-III antibodies in haematology staff. Lancet 1985;i:217.

[55] Krown SE, Real FX, Cunningham-Rundles S, et al. Preliminary observations on the effect of recombinant leukocyte A interferon in homosexual men with Kaposi's sarcoma. N Engl J Med 1983;308:1071–6.

[56] Anonymous. Progress on AIDS. FDA Drug Bull 1985;15:27–32.

[57] Mitsuya H, Popovic M, Yarchoan R, Matsushita S, Gallo RC, Broder S. Suramin protection of T cells in vitro against infectivity and cytopathic effect of HTLV-III. Science 1984;226:172–4.

[58] Broder S, Collins JM, Markham PD, et al. Effects of suramin on HTLV-III/LAV infection presenting as Kaposi's sarcoma or AIDS-related complex: clinical pharmacology and suppression of virus replication in vivo. Lancet 1985;ii:627–30.

[59] Chaisson RE, Allain J-P, Leuther M, Voberding PA. Significant changes in HIV antigen level in the serum of patients treated with azidothymidine. N Engl J Med 1986;315:1610–11.

[60] Volberding PA, Lagakos SW, Koch MA, et al. Zidovudine in asymptomatic human immunodeficiency virus infection. A controlled trial in persons with fewer than 500 CD4-positive cells per cubic millimeter. N Engl J Med 1990;322:941–9.

[61]   Ruskin J, LaRiviere M. Low-dose co-trimoxazole for prevention of *Pneumocystis carinii* pneumonia in human immunodeficiency virus disease. Lancet 1991;337:468–71.

[62]   Abdool Karim SSA. Nelson R Mandela (1918–2013). Science 2014;343:150.

[63]   Editorial. The toll from 3 deadly diseases. NY Times; October 12, 2013.

[64]   Anonymous. PEPFAR: a triumph of medical diplomacy. Science 2013;342:1466.

[65]   Goudsmit J. Joep Lange (1954–2014). Science 2014;345:881.

[66]   Brandt AM. How AIDS invented global health. N Engl J Med 2013;368:2149–52.

[67]   McNeil Jr. DG. Advocating pill, U.S. signals shift to prevent AIDS. NY Times 2014; May 15:A1.

[68]   Furtado MR, Callaway DS, Phair JP, et al. Persistence of HIV-1 transcription in peripheral-blood mononuclear cells in patients receiving potent antiretroviral therapy. N Engl J Med 1999;340:1614–22.

[69a] Walmsley SL, Antela A, Clumeck N, et al. Dolutegravir plus Abacavir-Lamivudine for the treatment of HIV-1 infection. N Engl J Med 2013;369:1807–18.

[69b] Hammer SM. Baby steps on the road to HIV eradication. N Engl J Med 2013;369:1855–7.

[70]   McLean AR. Coming to an airport near you. Science 2013;342:1330–1.

Chapter 5

Hemophilia: An Affinity for Blood

Greek haima (blood) + philia (to love) [1].

Total dependency on blood products set the stage for the catastrophic spread of acquired immunodeficiency syndrome (AIDS) among hemophiliacs. As noted in chapter 2, commercial clotting factor concentrates became widely available in the 1970s and were used by hemophiliacs at their homes, schools, and workplaces. The concentrates were effective in controlling hemorrhages, and they were also used to prevent bleeding. Pharmaceutical companies established extensive plasma collection facilities to feed the increasing demand for their products. Home health care companies arranged for delivery of the products to the home, submitted claims to insurance companies on behalf of patients, and paid the manufacturers. Everyone was satisfied with these arrangements— plasma donors received payment from the concentrate producers, manufacturers had a brisk market for their products, and the life span and quality of life for hemophiliacs were greatly improved. This "love" affair of hemophiliacs with clotting factor concentrates was shattered by the AIDS epidemic.

The first recognition that the disease was present in the hemophilia community came on Jul. 9, 1982, in a letter from William H. Foege, the Assistant Surgeon General, to all Hemophilia Treatment Centers [2]. He wrote that three men with hemophilia had developed *Pneumocystis jirovecii* pneumonia. This kind of pneumonia had previously been observed only in homosexual men, intravenous drug abusers, and Haitians who had recently entered the United States. Dr Foege indicated that physicians caring for hemophiliacs should promptly report patients with similar infections to their State Health Departments or the Centers for Disease Control (CDC). Detailed case histories of these three patients were published in the July 16 issue of the CDC's *Morbidity and Mortality Weekly Report* [3].

The first hemophiliac reported with AIDS was a 62-year-old man who became aware of weight loss in February 1981. In December, during hospitalization for elective knee surgery, he reported cough and fever; a chest radiograph

Linked by Blood: Hemophilia and AIDS. DOI: http://dx.doi.org/10.1016/B978-0-12-805302-7.00005-7

showed pneumonia. Because he failed to improve with treatment, a lung biopsy was performed, revealing *P. jirovecii*. He was treated with antimicrobial agents but died soon thereafter. The second individual, a 59-year-old man with hemophilia, noted gradual onset of weight loss, difficulty swallowing, mouth ulcers, and enlarged lymph glands under his jaw, all beginning in October 1980. In February 1982, he was diagnosed with oral thrush, a fungal infection of the mouth. Three months later he was hospitalized because of nausea, vomiting, and fever, and he was found to have pneumonia. A lung biopsy showed *P. jirovecii* as well as infection with cytomegalovirus. Despite treatment with antibiotics and antiparasitic drugs, he died within 2 months. The third patient was a 27-year-old hemophiliac who became ill in October 1981, with pneumonia involving both lungs. A lung biopsy showed *P. jirovecii*, and improvement came after a 3-week course of antibiotics. Four months later, he was treated for oral thrush, and 2 months after that he required hospitalization for fever, an enlarged spleen, anemia, and a low white blood cell count. After an extensive evaluation, he was found to have an atypical form of tuberculosis.

These three hemophiliacs fulfilled the criteria for the diagnosis of AIDS as defined by the CDC in 1982 [4]. They had diseases considered specific for AIDS or associated with AIDS: *P. jirovecii* pneumonia, candidiasis or thrush, cytomegalovirus disease, and tuberculosis. There were reports of other hemophiliacs with a variety of nonspecific symptoms that might be harbingers of AIDS, such as throat pain and difficulty swallowing (possibly indicating thrush), shortness of breath and persistent cough (features observed with *P. jirovecii* pneumonia), intractable diarrhea (consistent with infection due to a parasite called cryptosporidiosis), shingles (a viral infection of the skin and nerves), swollen lymph glands, and unexplained weight loss. The CDC became very concerned that AIDS might be spreading to the hemophilia community and indicated that physicians should report any type of persistent infection or organ dysfunction in their hemophilic patients.

All three of the hemophiliacs with AIDS had received long-term treatment with factor VIII concentrates. When they became ill, extensive laboratory testing showed that they had severely depleted immune systems. Although none was a homosexual or intravenous drug abuser, they all had clinical and laboratory features that were strikingly similar to those encountered in the more than 400 AIDS patients referred to the CDC between 1978 and 1982. Richard A. Moates of the Food and Drug Administration (FDA) wrote, "The occurrence of pneumocystis pneumonia in these three patients would add some support to the theory that an unknown infectious agent other than the known hepatitis viruses might be present in AHF[1] and be involved in suppressing the immune system, thus allowing opportunistic infections to occur" [5]. He noted that the FDA's Office of Biologics was collaborating with the National Institutes of Health

---

1. AHF, anti-hemophilic factor, was the name for factor VIII concentrates in the early 1980s.

(NIH) to define the nature and mechanism of the immunodeficiency and the frequency of infection. He further indicated that the FDA was looking into questions raised about the safety of blood products.

In response to the CDC alert, Charles J. Carman, President of the National Hemophilia Foundation (NHF), and Louis Aledort, Medical Co-Director, met with Edward N. Brandt, Jr, the US Assistant Secretary for Health [6]. They agreed to begin surveillance of hemophilia patients, collect evidence to determine whether the illness in the three hemophilic patients was part of a pattern, and attempt to assess the safety of blood products. Secretary Brandt approved appointments of members of the NHF to a special panel established by the Department of Health and Human Services to study the problem of AIDS.

The Medical and Scientific Advisory Council (MASAC) of the NHF met on October 2, 1982, and because of their concern about the new disease, they invited Dr Bruce Evatt of the CDC to update the Committee on the state of knowledge regarding AIDS. Evatt reviewed the information that the CDC had accumulated and noted that the working group on AIDS suspected that they were seeing a new virus or new epidemic [7]. Furthermore, he reported that researchers had observed viral particles in clotting factor concentrates, but their nature could not be more specifically defined. Despite Dr Evatt's concern that commercial concentrates might be contaminated by a virus, MASAC was not convinced that the use of these blood products should be curtailed [8]. The NHF's physician advisors weighed the risks of what appeared to be an "exotic" malady affecting very few hemophiliacs against the clear benefits of blood products. They felt that there was insufficient evidence to recommend any changes in hemophilia treatment. In September 1982, NHF disseminated an informational issue of their newsletter, Hemophilia NewsNotes, entitled *Acquired Immune Deficiency Syndrome: Questions and Answers* [9]. One of the questions was "should the hemophiliac change or stop his treatment with factor VIII or factor IX?", and the answer was "At the present time, there is no specific evidence to warrant changing the use of factor VIII or factor IX. By all means, do not modify treatment on your own. Any questions or concerns you may have over this issue should be communicated to your physician or treatment center." However, MASAC did urge that manufacturers of source plasma used for factor VIII products should exclude from plasma donation all persons belonging to groups at high risk for AIDS (homosexuals, intravenous drug abusers, and immigrants of Haitian background).

Although AIDS had been reported in only a few hemophiliacs, some physicians were suspicious that the clotting factor concentrate was the source of the infection, and Dr Evatt's comments about viral contamination reinforced this view. But even assuming that the concentrates did infect hemophiliacs, other physicians argued that there probably wouldn't be serious consequences of such an infection. They cited the example of hepatitis viruses; although these viruses had been extensively transmitted by blood products, very few hemophiliacs showed signs of severe liver disease. Apparently, these physicians were unaware that the new infection was actually much more severe than hepatitis. In October 1982, at the time of the MASAC recommendations, the CDC reported that 691 persons (mainly homosexuals) had been diagnosed with AIDS and 278 (40%) died. If this same mortality rate occurred in hemophiliacs, then the risk of dying from AIDS would be far greater than any of the other risks encountered by people with bleeding disorders, such as getting severe liver disease or having a fatal, untreated hemorrhage.

By December 1982, an NHF Chapter Advisory noted that the number of individuals with hemophilia affected by the new disease had increased to eight and was strongly suspected in two more [10]. Despite this report, NHF was still uncertain whether clotting factor concentrates were the source of the disease [11] and recommended that patients already using concentrates should continue their use; however, they did suggest that concentrates should not be introduced to children from birth to age 4 years, newly diagnosed patients, or those with mild disease. NHF also called for commercial and voluntary blood banks to develop more restrictive blood donor criteria that would screen-out high-risk groups such as paid plasma donors. However, studies showed that the effect on the hemophilic immune system was the same whether the plasma used for concentrate preparation came from volunteer donors or paid donors [12]. The country of origin of the plasma was the best predictor of infection risk. Hemophiliacs exposed to concentrates prepared from Scottish or Danish plasma had less evidence of exposure to HTLV-III than those treated with concentrates derived from US plasma (6.7% vs 39.7%) [13].

Because of the concern about the safety of clotting factor concentrates prepared from paid donor plasma, some Hemophilia Centers recommended that their patients should switch from commercial brands to concentrate prepared by the Red Cross because that agency used plasma from volunteer donors exclusively [14]. However, even the Red Cross concentrate was made by pooling the plasma from many donors, and homosexuals constituted a notable proportion of the volunteer plasma donors. Infected plasma from these donors could enter the pools from which the clotting factor concentrates were made. Although Hemophilia Center Directors suspected that AIDS was transmitted by blood products, they were reluctant to recommend any change in concentrate usage, fearing that the consequences would be inadequately treated hemorrhages, development of crippling joint deformities, and increased hospitalizations.

Carol Kasper and colleagues at the Los Angeles Orthopedic Hospital, which provided care to many patients with hemophilia, issued an informational bulletin about AIDS in January 1983 [15]. They noted that a diagnosis of AIDS required the presence of either Kaposi's sarcoma or an infection consistent with a defect in immunity, termed an opportunistic infection. More than a dozen types of microbes were the cause of such infections; the organism most frequently cited was *P. jirovecii*. Most ill persons resided in New York or San Francisco, but the disease was also reported in Europe and Haiti. As of November 1982, homosexual or bisexual men comprised 74.5% of the cases, intravenous drug users comprised 14.1%, Haitians comprised 5.7%, and hemophiliacs comprised only

0.7%. There were suggestions that the mysterious agent, probably a virus, originated in Haiti, a popular and cheap vacation resort for New Yorkers. The virus was disseminated among the more promiscuous members of the homosexual community in New York and other large cities.

Although the number of hemophiliacs with AIDS was small (only 10 at the time of this report in January 1983), Kasper and colleagues [15] reported that a substantial percentage of persons with hemophilia had changes in their immune cells similar to those seen in patients with AIDS. This observation was based on a study showing a striking reduction in the ratio of helper (CD4) to suppressor (CD8) lymphocytes in 9 of 12 apparently healthy hemophiliacs regularly treated with concentrate [16]. It was unclear whether these immunologic findings were characteristic of hemophiliacs in general, associated with the use of clotting factor concentrates, or due to infection by a virus [17]. To try to resolve this question, researchers from the University of Pittsburgh studied 18 men with hemophilia and their spouses, and 19 hemophilic children and their siblings [18]. Evidence of immune suppression was found in the hemophiliacs but not in their wives, brothers, or sisters, leading the investigators to conclude that the changes in immune cells were treatment-related and not due to genetic factors. It seemed plausible that the impaired immunity in these hemophilic men and boys was due to an infectious agent that was not easily transmitted to family members, and that these hemophiliacs might already be on their way to developing AIDS.

These reports raised the possibility that persons with hemophilia could be harboring the AIDS agent but could still be without symptoms. Husbands asked whether they should continue to have sexual relations with their wives or kiss their children. Nurses and laboratory workers were concerned with protecting themselves against infection when examining patients and drawing blood samples. Precautions were clearly indicated when dealing with AIDS patients, but whether similar safeguards were needed for hemophiliacs without symptoms was uncertain. Persons with hemophilia and their families became apprehensive because they recognized that their physicians were unsure about the appropriate course of action.

The NHF asked its Mental Health Committee to address the intense stress affecting the hemophilia population. The Committee summarized the following patient concerns [19]:

- Treatment with clotting factor concentrate was viewed by some hemophiliacs as a form of Russian roulette because there was no way of knowing if it would transmit AIDS.
- Others felt that limiting home use of concentrate represented a return to helplessness and dependence.
- Some found the linkage with homosexuality and intravenous drug abuse repugnant.
- Many developed a "leper-like" self-image because of the possible risk of transmitting AIDS to family and friends.

The Mental Health Committee recommended that medical staff should provide regular opportunities for discussing updated information with patients and their families. Hemophilia Center staff should be alert to signs of mental distress and offer treatment and psychiatric consultation as necessary. In the face of uncertainty, shared decision-making between the patient and physician was strongly advocated. Finally, they encouraged health care providers to examine their own feeling about AIDS and to avoid using denial in presenting overly optimistic views to their patients.

Kasper and colleagues [20] suggested that exposure to the AIDS agent might be reduced by limiting the overall use of blood products and switching from concentrate to cryoprecipitate obtained from selected donors. There was scientific support for the suggestion that cryoprecipitate might be safer than the concentrate manufactured by pharmaceutical companies. A study of 24 Finnish hemophiliacs treated exclusively with their own locally prepared cryoprecipitate showed that none had the immunologic abnormalities commonly observed in US patients treated with concentrate [21]. As early as 1980, a procedure was described that enabled one adult to provide sufficient cryoprecipitate for the needs of a young child [22]. The donor, usually the father of a child with hemophilia, was bled and the cryoprecipitate was harvested from his plasma. Then, the red blood cells and remaining plasma were returned to the donor. By use of this procedure, five children could be supported by a single parental donor for periods from 6 months to 4 years [23]. The great advantage of this method was that the child was exposed to the blood of only one donor, and thus could almost certainly avoid infection by either hepatitis or HIV/AIDS. However, few blood banks and hospitals adopted this approach because it required a dedicated staff, special equipment, designated donors, and increased costs. Had they been offered this option, there is little doubt that most families of hemophiliacs would have enthusiastically embraced it, despite the discomforts and time commitments that the procedure required.

Most Hemophilia Centers found treatment with cryoprecipitate to be impractical for patients with frequent hemorrhages, especially for those using home treatment [24]. It was not always easy to obtain cryoprecipitate; its availability depended on a blood bank willing to prepare and stock the material. The bags of cryoprecipitate occupied a large amount of space in the home freezer and required special insulated containers if transported elsewhere. Thawing and dissolving the material was tedious, and often particulate material clogged the small filters needed for infusions. For these and other reasons, cryoprecipitate was not a popular choice for home therapy, and commercially prepared clotting factor concentrates continued to be preferred by most patients and their caregivers despite their concerns about product safety. In fact, in their question and answer booklet, the NHF responded to the question "Is there an increased risk of developing AIDS from the use of concentrate as compared to the use of cryoprecipitate or fresh frozen plasma?" with "To date there is no specific evidence supporting any greater risk with concentrate than with cryoprecipitate or fresh frozen plasma. We are currently studying this matter in an intensive fashion" [9].

The answer given by the NHF to this question seems misleading because there were probably many in the leadership that had a strong suspicion that concentrate was more risky than cryoprecipitate or plasma; however, the NHF response was in line with their official recommendation that hemophiliacs already using concentrate should continue to do so. However, NHF did advise that newly diagnosed hemophiliacs should be given cryoprecipitate. This recommendation was influenced by a study published in January 1983 that examined immunologic markers in two groups of apparently healthy hemophiliacs: 11 treated solely with concentrate and 8 treated only with cryoprecipitate [25]. The concentrate group had a relative decrease in CD4 cells and impaired responses to immune stimulation, whereas the numbers and function of CD4 cells in the cryoprecipitate group were similar to those in persons without hemophilia. Abnormal results were observed in more concentrate-treated than cryoprecipitate-treated hemophiliacs (73% vs 25%). A second study published in the same journal was less definitive; it reported only slightly lower numbers of CD4 cells in users of concentrate as compared to users of cryoprecipitate (44% vs 49%) [26]. Other investigations found that cryoprecipitate was not innocuous; for example, some reported immunologic abnormalities similar to those observed in concentrate users [27a,b]. Another study [28] suggested that it was the intensity of treatment rather than the type of product that altered the immune system; concentrate was usually infused more frequently and in greater amounts than cryoprecipitate; therefore, it showed more adverse effects on immunity.

These observations left hemophiliacs and their caregivers in a quandary. Clotting factor concentrates were effective but clearly a suspect. Cryoprecipitate was possibly safer but not always available, and it was less effective than concentrates. Furthermore, the prevalence of hepatitis markers was the same for users of cryoprecipitate and users of concentrate, suggesting that AIDS transmission would be similar with the two products [29]. Patients and families were under extreme stress; an NHF Medical Bulletin described "the makings of a nightmare—a lethal threat from a mysterious source" [14] and "(Clotting factor) concentrate that was previously received as lifesaving now has the implication of a poison." Many hemophiliacs used smaller doses or delayed treatment for joint bleeds. However, this strategy only led to more severe hemorrhages and the need for larger doses of concentrate to control bleeding. To help physicians cope with the challenges posed by the new syndrome, Hemophilia Centers in Ohio and Texas organized symposia that included presentations on testing for immunodeficiency and the status of new blood products. The speakers reported that safer clotting factor concentrates were in development, but they had little else to offer for the existing management of persons with severe hemophilia or those with AIDS.

In May 1983, the Medical Advisory Board of the World Federation of Hemophilia distributed background material on AIDS prepared by its

Chairperson, Shelby L. Dietrich. After reviewing the history of the epidemic, she listed the most widely accepted theories about its causes [30]. Foremost was the notion that acquired immune deficiency was due to a heavy bombardment of the immune system by foreign proteins. It was suspected that homosexual practices enabled semen and sperm to penetrate the fragile lining of organs such as the bowel, then enter blood and lymph channels, and, finally, overstimulate the immune system. Intravenous drug injectors were exposing their immune systems to dirty needles and syringes containing blood and other materials. Hemophiliacs were thought to be at risk because they received frequent intravenous infusions of impure clotting proteins that might be blunting their immune responses. The second most popular theory (which eventually proved to be correct) was that a transmissible agent, such as a virus, was present in the body fluids of homosexuals as well as in the blood that they donated. Others thought that both theories were tenable.

Because of the possibility of acquiring AIDS through blood transfusion, there was intense concern about donation of blood or plasma by persons belonging to high-risk groups (homosexuals and drug abusers). Dr Dietrich noted that the pharmaceutical industry had been quite cooperative in attempts to exclude plasma donors at high risk for AIDS. They provided educational material to donors, asked questions designed to identify and exclude high-risk donors, and checked for early signs of AIDS, such as unexplained weight loss and swollen glands. Persons with symptoms of AIDS, those who considered themselves at risk for AIDS, and those who had intimate contact with someone who may have had AIDS were asked to refrain from donating blood.

Dr Dietrich also recognized that the possibility of acquiring AIDS from blood components continued despite efforts to improve the donor pool. She wrote that the "treatment of hemophilia rests basically in replacement of the missing coagulation factors from human source blood or plasma products." She summarized the dilemma faced by patients with severe hemophilia, shown here in modified form:

Type of treatment	Risks	Benefits
None or minimal use of blood, plasma, cryoprecipitate, or concentrates	Death or shortened life span, severe pain, and disability	Cost: none to low
	Invalid lifestyle, dependency, nonproductive	Minimal exposure to hepatitis and AIDS
Early, intensive treatment or preventative therapy with concentrates or cryoprecipitate	High cost	Normal lifestyle
	Large exposure to blood products	Normal vocational and professional opportunities
	Increase in hepatitis, cirrhosis, AIDS	Prevention of crippling, able to become a productive member of society

She concluded that physicians and patients together should "weigh the relative benefits/assets and risks/problems of various modes of treatment and reach a decision beset with ambiguities and uncertainties."

The NHF, however, unambiguously urged that clotting factor use should be maintained [31]. They recognized that media coverage of AIDS and a new report that a brand of concentrate was being recalled because one of the plasma donors developed symptoms of AIDS, were causing some patients to abandon the use of blood products. They urged patients and physicians to carefully evaluate blood products and stated that recall actions *should not cause anxiety or changes in treatment programs* (underlined in the Bulletin). They emphasized that "the life and health of hemophiliacs depends upon blood products" and that only a small number (12 at the time) of the nation's 20,000 hemophiliacs had developed AIDS. However, had they broadened their perspective, they would have observed that AIDS was becoming rampant in other populations: 1366 cases of AIDS and 520 (38%) deaths had already occurred among homosexuals and drug abusers.

The Blood Products Advisory Committee of the Office of Biologics became increasingly concerned about the safety of the clotting factor concentrates. John C. Petricciani, the Director, convened a conference of blood bankers, concentrate manufacturers, and the NHF in July 1983 [32]. In his introductory remarks, Petricciani called on all establishments collecting source plasma to discourage blood donations from persons at risk for AIDS and recommended that donor screening should include questions about the symptoms and signs of AIDS as well as an examination for evidence of recent weight loss or enlarged lymph glands. But he also noted that there was a difference between signs and symptoms that could be compatible with AIDS and the actual clinical diagnosis of this syndrome. He warned Committee members that a mandatory recall of products prepared from suspect donors could have a major impact on the supply of therapeutic materials.

There followed presentations by the NHF, industry representatives, and the CDC. The position of the NHF was presented by MASAC's Louis Aledort. The NHF recommendation was that a specific lot of concentrate should be recalled if it included plasma from a patient with AIDS or with characteristics suggestive of AIDS. Michael Rodell of Armour Laboratories represented the Pharmaceutical Manufacturers Association (PMA). He provided details about the plasma requirements for concentrate production, noting that from 1000 to 10,000L (quarts) of plasma were required to produce enough concentrate to treat 12–125 hemophilia patients for a year. He estimated that frequent plasma donors made 40–60 donations per year. If a particular donor was found to have AIDS and if a decision was made to recall all concentrate prepared from that donor's plasma collected during the previous year, then 25–250 million clotting factor units would need to be withdrawn. This would have a substantial impact on the concentrate supply. Steven J. Ojala of Cutter Laboratories said that the PMA recommended against mandatory recalls and instead suggested continuing its established screening policies and discarding plasma from suspect donors. Recalls could be made on a case-by-case basis in close consultation with the FDA.

Bruce Evatt spoke on behalf of the CDC's task force on AIDS. He reported that every hemophiliac diagnosed with AIDS had received clotting factor concentrates. In addition, 17 other patients with AIDS had been identified, and their only risk factor appeared to be transfusion of blood or blood products. Despite this presentation, Committee members still felt that the risk of AIDS from concentrates or transfusion was not established. They discussed whether there should be a mandatory recall of blood products if a plasma donor was identified as an AIDS patient or had signs and symptoms of AIDS. They learned that one manufacturer (Hyland Therapeutics Division) had recalled 187 vials of concentrate when a plasma donor was found to have symptoms of AIDS. This showed that the industry was willing to take voluntary action. The Committee members concluded that in the absence of comprehensive data on risk, and considering the potential for serious disruption of the supply of hemophilia care products, the FDA should not adopt a policy of mandatory recall. This failure to take regulatory action allowed continued marketing of contaminated concentrate and the potential infection of more hemophiliacs.

In the month following this meeting (August 1983), Hyland and the Red Cross recalled additional lots of concentrate that had been made using plasma from a donor who was confirmed to have died of AIDS 10 months after the donation. It was apparent that donors could appear well but still be harboring the disease. In fact, the CDC had reported as early as April 1982 that AIDS had an incubation period of at least 10.5 months [33]. Therefore, a person could be infected but not display the characteristic features of the syndrome for months or years. Donors and recipients could appear healthy but could be incubating the virus. The 17 cases described in hemophiliacs up to that time were just the tip of the iceberg!

KEY POINTS

- AIDS was first reported in hemophiliacs in early 1982, and it was suspected that blood products were the source because of similarities to infection by hepatitis viruses in the 1970s.
- However, NHF and its scientific advisory committee considered that the risk/benefit ratio favored undiminished use of blood products because the benefit of controlling bleeding was clear, but the risks of the new infection were unknown.
- The chance of infection from commercially prepared clotting factor concentrates was considerable because thousands of liters/quarts of plasma were required for plasma fractionation, and a single donation containing the virus could contaminate the entire lot. The risk with cryoprecipitate was much less, but cryoprecipitate was more difficult to store and infuse.

- Apparently healthy men and boys with hemophilia had decreases in CD4 T cells and other immunologic abnormalities, and these were observed more often in those treated with concentrate than with cryoprecipitate; these changes were not observed in their wives, brothers, or sisters.
- NHF advised that cryoprecipitate was preferred for hemophiliacs not previously exposed to concentrates, and that manufacturers should try to exclude persons at high risk for AIDS from the donor pool.
- At a meeting in 1983, the consensus of blood bankers and the FDA was that there should be no mandatory recall of concentrate even if one of the donors was found to have AIDS.

REFERENCES

[1] Online Etymology Dictionary. Accessed February 2014.
[2] Foege WH. *Pneumocystis carinii* pneumonia among patients with hemophilia: Centers for Disease Control, Department of Health and Human Services; July 9, 1982.
[3] Centers for Disease Control. *Pneumocystis carinii* pneumonia among patients with hemophilia. MMWR Morb Mortal Wkly Rep 1982;31:365–7.
[4] National Hemophilia Foundation and Centers for Disease Control Survey for Acquired Immune Deficiency Syndrome (AIDS) among patients attending hemophilia treatment centers (HTCs). Spectrum of disease presentation in the Acquired Immune Deficiency Syndrome (AIDS); 1982.
[5] Moates RA. *Pneumocystis pneumonia* in patients with hemophilia. A talk paper issued by the FDA on July 23, 1982.
[6] Hemophilia Newsnotes, Medical Bulletin #2, Acquired immune deficiency syndrome (AIDS); July 30, 1982.
[7] Resnik S. Blood saga. Berkeley, CA: University of California Press; 1999, p. 117.
[8] Minutes of the October 2, 1982 MASAC Meeting.
[9] Hemophilia Newsnotes. Acquired immune deficiency syndrome: questions and answers; September 1982.
[10] Hemophilia Newsnotes, Chapter advisory #4, Acquired immune deficiency syndrome (AIDS) update; December 9, 1982.
[11] Hemophilia Newsnotes, Medical Bulletin #4, Chapter advisory #5. AIDS: implications regarding blood product use; December 21, 1982.
[12] Cable RG, Hoyer LW, Marchesi S, Mukherji B, Morse EE, Saxton P. Influence of plasma source on T-lymphocyte subpopulations in hemophiliacs using factor VIII concentrate. N Engl J Med 1983;309:1057–8.
[13] Melbye M, Madhok R, Sarin PS, et al. HTLV-III seropositivity in European haemophiliacs exposed to factor VIII concentrate imported from the USA. Lancet 1984;ii:1444–6.
[14] Weiss AE. Regional Comprehensive Hemophilia Center of Central & Northern Illinois Newsletter; January 1983.
[15] Kasper CK, Dietrich SL, Boylen AL, Ewing NP. Acquired immune deficiency (AIDS) and hemophilia. Hemophilia Bull, January 1983:1–4.
[16] Goldsmith JC, Moseley PL, Monick M, Brady M, Hunninghake GW. T-lymphocyte subpopulation abnormalities in apparently healthy patients with hemophilia. Ann Intern Med 1983;98:294–6.
[17] Aledort LM. AIDS: an update. Hosp Pract 1983;18(9):159–71.

[18] Ragni MV, Bontempo FA, Lewis JH, Spero JA, Rabin BS. An immunologic study of spouses and siblings of asymptomatic hemophiliacs. Blood 1983;62:1297–9.

[19] Agle DP for the Mental Health Committee. Mental health aspects of AIDS. Hemophilia Newsnotes, Medical Bulletin #6; March 9, 1983.

[20] Kasper CK. How do you cope with AIDS? Hemophilia Bull November 1983.

[21] Rasi VPO, Koistinen JLK, Lohman CM, Silvennoinen OJ. Normal T-cell subset ratios in patients with severe haemophilia A treated with cryoprecipitate. Lancet 1984;i:461.

[22] McLeod BC, Cole ER, Sassetti RJ, Pierce MI. Factor VIII collection by pheresis. Lancet 1980;ii:671–3.

[23] McLeod BC, Sassetti RJ, Cole ER, Scott JP. A high-potency, single-donor cryoprecipitate of known factor VIII content dispensed in vials. Ann Intern Med 1987;106:35–40.

[24] Kasper CK. What plasma product should be used? Hemophilia Bull April 1984, p. 1–2.

[25] Lederman MM, Ratnoff OD, Scillian JJ, Jones PK, Schacter B. Impaired cell-mediated immunity in patients with classic hemophilia. N Engl J Med 1983;308:79–83.

[26] Menitove JE, Aster RH, Casper JT, et al. T-lymphocyte subpopulations in patients with classic hemophilia treated with cryoprecipitate and lyophilized concentrates. N Engl J Med 1983;308:83–6.

[27a] Kessler CM, Schulof RS, Goldstein AL, et al. Abnormal T-lymphocyte subpopulations associated with transfusions of blood-derived products. Lancet 1983;i:991–2.

[27b] Ludlam CA, Carr R, Veitch SE, Steel CM. Disordered immune regulation in haemophiliacs not exposed to commercial factor VIII. Lancet 1983;i:1226.

[28] Lee CA, Janossy G, Ashley J, Kernoff PBA. Plasma fractionation methods and T-cell subsets in haemophilia. Lancet 1983;ii:158–9.

[29] Levine PH. T-cell subsets in hemophilia. N Engl J Med 1983;308:1293.

[30] Dietrich SL. Background material on AIDS for the Medical Advisory Board of the World Federation of Hemophilia; May 1983.

[31] Hemophilia Newsnotes, Medical Bulletin #7, Chapter Advisory #8. NHF urges clotting factor use be maintained; May 11, 1983.

[32] Council of Community Centers Newsletter; July 15/22, 1983.

[33] Shilts R. And the band played on. New York, NY: St Martin's Press; 1987, p. 147.

Chapter 6

Blood as a Vehicle for the Spread of AIDS

In October 1982, the Centers for Disease Control (CDC) received a report of a 20-month-old infant from the San Francisco area who died of a severe acquired immunodeficiency syndrome (AIDS)-like illness [1]. This baby had required multiple blood transfusions soon after birth because of jaundice and remained in the hospital for 8 weeks; during that time, he received blood products from 18 donors [2]. He then appeared well until 4 months of age, when enlargement of the liver and spleen were detected, as well as recurrent ear infections, oral thrush, and a skin rash. In addition, he lost weight and had developmental delays. At 10 months, he began to have vomiting and diarrhea and became jaundiced. His blood counts were low and there was evidence of immune deficiency; at age 18 months, he was found to be infected by an atypical tuberculosis organism. At this point, AIDS was suspected and investigations revealed that one of the donors who provided blood for the child had died of AIDS several months after the donation, suggesting that the blood of this donor might have transmitted the disease to the infant. There had been much speculation about the cause of AIDS, but until this report there had been no firm evidence that a communicable agent was involved. The infection in this child showed that you did not have to be a homosexual or drug abuser to get AIDS; even babies were vulnerable. If the disease was transmissible by blood, then large segments of the general population could be at risk.

This concern about the risk from blood received support from a second case of suspected transfusion-associated AIDS [3]. A 53-year-old man had coronary artery bypass surgery in November 1980. On the day of surgery, he received blood products from 16 donors. After hospital discharge, he experienced weight loss, fatigue, and fever; 4 months later, he was rehospitalized for suspected pneumonia. A lung biopsy revealed *Pneumocystis jirovecii* infection. Laboratory evaluation showed a profound decrease in CD4 T-cells. The patient was not a homosexual or intravenous drug abuser, and his wife had no evidence of infection or immunodeficiency. The authors of this report believed that their patient had contracted AIDS from blood and blood product transfusions.

Linked by Blood: Hemophilia and AIDS. DOI: http://dx.doi.org/10.1016/B978-0-12-805302-7.00006-9

This communication and the continuing reports of AIDS in hemophiliacs receiving blood and blood products led the CDC to convene a workshop on January 4, 1983 to discuss ways of halting the spread of this new disease [4]. Participants included blood bank directors, government officials, and a representative of the National Hemophilia Foundation (NHF). Although it was suspected that AIDS might be transmitted by blood, it was noted that only a tiny fraction of those receiving transfusions had developed the disease. But a CDC official warned that there might be a lag period of a year or more between the time of exposure to the causative agent and the onset of symptoms, implying that the number of people infected might be much larger than the number with overt disease. Despite this possibility, most conference participants were unwilling to make recommendations about donor screening or blood testing. A leading blood bank director was concerned that publication of single case reports linking infections with transfusion would create anxiety and fear in the public [5], and the NHF spokesperson said that it was too early to conclude that a transmissible agent contaminated blood products.

Nevertheless, soon after this workshop, the NHF requested that immediate steps should be taken to remove homosexuals and others at risk for AIDS from the donor pool because of the possibility that their blood might transmit the infection. This action was opposed by the leading blood collection groups (the Red Cross, American Association of Blood Banks, Community Council of Blood Centers) because of their reluctance to question potential donors about their sexual orientation [6]. However, in March 1983, they acceded to a recommendation from the Food and Drug Administration (FDA) that they institute procedures to inform members of groups at high risk for AIDS to refrain from donating blood. Although the Blood Bank group repeatedly stated that the risk of contracting AIDS by transfusion was only one in a million patients who underwent transfusion, by August 1983 the CDC had reports of 18 persons who developed AIDS with no risk factors other than a history of transfusions. Further investigation of seven of these patients showed that all had received blood from at least one high-risk donor, clearly incriminating transfusion as a risk factor for AIDS [7]. Calculations performed in January 1984 showed that the number of cases of AIDS in patients receiving transfusions exceeded the number expected if transfusion were not a risk factor by almost 6 to 1 [8]. A year later, the CDC discovered that 90 AIDS patients had donated blood within the previous 5 years and established that 53 of the 4690 AIDS cases then extant were transfusion-related [9].

Could blood itself, especially the transfusion of more than one pint, increase the susceptibility of recipients to a variety of infections? A study of more than 50 patients who underwent multiple transfusions reported evidence of T-cell activation and decreases in some lymphocyte populations, but the levels of CD4 T-cells were normal, a finding inconsistent with HIV infection [10]. Furthermore, none had evidence of unusual infections or other clinical features of AIDS. The investigators concluded that the immunologic profile of patients with chronic transfusions differed from that observed in homosexuals and hemophiliacs, thereby making it very unlikely that uncontaminated blood was a cause of AIDS.

Transfusion of platelets as well as blood could induce the disease. A woman was given 6 units of platelets because of a low platelet count, and she developed AIDS 8 months later [11]. She had never received red cell transfusions and was not in any of the known AIDS risk groups. However, one of the platelet donors was a homosexual man with enlarged lymph glands and an estimated 500 sexual partners in the previous 5 years; he was presumably the source of the infection. This and other reports provided strong evidence that blood and blood products, probably contaminated by a virus, were the source of the illness.

An outspoken skeptic of the theory positing a specific AIDS virus was Peter H. Duesberg, a professor of molecular and cellular biology at the University of California, Berkeley. He argued that the microbe identified in persons developing AIDS was just a harmless passenger virus. He noted that the percentage of symptomatic carriers of this virus was low, the latent period for AIDS was long, and no viral gene had ever been identified that would induce susceptibility to other infections [12]. However, these observations do not exclude human immunodeficiency virus (HIV) as the cause of AIDS; they only indicate that it has unique characteristics that distinguish it from other viruses. Duesberg also suggested that the immune deficiency characteristic of AIDS was due to contaminants other than HIV in blood or clotting factor concentrates [13]. But the concept that HIV was central to the development of AIDS was supported by a study that showed a direct relationship between HIV infection in donors and the appearance of AIDS in the recipients, and the observation that all blood products (red cells, plasma, or platelets) prepared from donors known to have previously transmitted AIDS to recipients conveyed the disease to all future recipients [14]. Furthermore, transmission of infection was most likely if the donor was within 23 months of an AIDS diagnosis, when viral levels were most elevated. In addition, deaths due to AIDS in hemophiliacs were associated with the receipt of HIV-positive, but not HIV-negative, factor VIII concentrate and occurred irrespective of the number of units of factor VIII administered [15]. This was convincing evidence that it was not the clotting factor, but rather HIV contaminating the clotting factor that was responsible for the disease.

This controversy about the cause of AIDS affected public policy; for example, Duesberg's ideas had a strong influence on Thabo Mbeki, the President of South Africa from 1999 to 2008, and led to a long delay in the distribution of antiretroviral drugs in that country. It has been estimated that the failure to provide anti-HIV drugs between 2000 and 2005 indirectly contributed to the death of 330,000 people and the mother-to-child transmission of HIV to 35,000 babies [16].

In the early 1980s, various procedures were instituted in an effort to safeguard the blood supply. For example, the South Florida Blood Service gave all prospective donors a card listing the groups of people at highest risk for AIDS and asked donors who belonged to one of these groups not to donate [17].

However, if such persons still wished to donate (perhaps because they were participants in a school or employee blood drive and did not want their coworkers to know they were in a special category), then they should dial a "hotline" within 4 hours of the donation to request that their blood not be used for transfusion, and no questions would be asked. This type of passive screening depended on the donor acknowledging membership in a high-risk group. However, there was the possibility that some people might not be in such a group but still harbor the virus, and occasionally donors might not even be aware that they were infected.

Most blood bankers recognized that this method of self-deferral was not very reliable; in fact, some gay persons thought that by donating blood they could overcome their personal fears of having AIDS [18]. A better way to exclude infected donors is by examination of the donor's blood for evidence of HIV infection. This could be done by measurement of the T-helper to T-suppressor cell ratio; reversal of this ratio is characteristic of the immune deficiency associated with the virus and would disqualify the donor. Because performing this test was time-consuming and expensive, it was not implemented by most blood banks; only the bank at Stanford University Hospital adopted it as a routine measure. Another indirect method to screen blood for AIDS is to search for markers of hepatitis B, because evidence of hepatitis is frequently observed in the same groups that are also at risk for AIDS: homosexuals, intravenous drug abusers, and prisoners. Cutter laboratories and some San Francisco Bay Area blood centers announced in April 1984 that they would begin screening all their donors for exposure to hepatitis B, although they recognized that a positive test result was not necessarily indicative of AIDS or future development of the disease [19].

Patients requiring blood transfusions became very concerned about the risk of getting AIDS. Many requested "directed donation": they wanted to receive only blood that had been donated specifically for them by family members or friends. They believed that such blood would be safer than blood obtained from random donors. Directed donations were strongly discouraged by blood banks for a number of reasons [20,21]. First, there was no proof that directed donations were any safer than blood donated by altruistic volunteers. Furthermore, by requiring blood banks to segregate donated blood into a special "designated" pool, the notion that designated blood was less likely to transmit disease was given credence [22]. Second, designated donations might be more risky because donors would feel obligated to donate, even if they were ill or in a high-risk group [23]. For example, a patient needing cancer surgery would be informed that 3–4 units of blood might be needed during the procedure. If the patient wanted directed donations, he/she would ask siblings, cousins, and close friends to donate blood on his/her behalf prior to surgery. One or more of these potential donors might have injected drugs in the past or be a homosexual, but they might find it difficult or embarrassing to admit to these practices and would proceed with the blood donation. Third, some directed donors might be new donors, and blood from new donors, as compared to repeat donors, was found to have two- to three-times the risk of disease transmission [24]. Finally, segregating directed from ordinary donations required additional processing, separate storage, and special dispensing, and these logistical steps increased the likelihood of errors. Promulgation of guidelines for prospective donors led to a decrease in the numbers of high-risk donors and stiffened resistance to directed donations. Opponents of directed donation were able to defeat a Florida Bill containing provisions to require blood centers to accept directed donations.

Public recognition that AIDS was associated with transfusions led to not only great trepidation about receiving blood products but also refusal to donate blood. People were afraid that blood donation equipment might be contaminated and that blood bank personnel were contagious. Margaret Heckler, the Secretary of the Department of Health and Human Services, donated blood herself to assure the public that there was no risk of AIDS to blood donors, and she encouraged healthy individuals to continue donating blood.

The hysteria about AIDS became so acute that patients with known AIDS, as well as homosexuals and hemophiliacs, became societal outcasts. Especially egregious was the reluctance of doctors, nurses, teachers, and officials to make physical contact with AIDS patients. A striking example of such discrimination was exemplified by the experience of Ryan White [25]. In December 1984, this 14-year-old boy with hemophilia developed pneumonia and was diagnosed with AIDS. When he was sufficiently recovered, he attempted to return to school but was denied readmission because the school board considered him to be a danger to his teachers and other students. After a lengthy appeals process, Ryan was eventually allowed to resume classes, but he was required to eat with disposable utensils, use separate bathrooms, and not participate in gym classes. This was despite an authoritative medical report that the risk of infection with AIDS was "minimal to nonexistent" in persons living in close but nonsexual contact with people with AIDS [26]. Even after he was permitted to return to school, he was shunned by his schoolmates and subject to disparaging remarks and threats. However, Ryan was not afraid to speak out about his life with AIDS, and he became a public spokesperson for the disease. He described the discrimination he faced when he tried to return to school, but he noted that eventually education about AIDS led to his acceptance by the community. He advocated for AIDS research and rejected criticism of homosexuality. As medical information accrued about the cause of AIDS and the public became better informed, people recognized Ryan's bravery and courage. Soon after his death at age 19, Congress enacted the Ryan White HIV/AIDS Program to provide qualified applicants with health care coverage and financial resources for coping with HIV [27].

The development of an antibody test for HTLV-III was a major breakthrough in the detection of infected blood donors. The US Public Health Service issued recommendations for blood and plasma screening [28], and the FDA licensed the antibody tests of two manufacturers in March 1985 [29]. Intensive efforts to discourage homosexual men from donating blood were implemented and extended to promiscuous heterosexual men and women, particularly prostitutes [30]. Donors were informed that their blood or plasma would be tested for the antibody; they were told not to donate if they did not want such testing. Once donors were found to have a positive test result, they were placed on a donor deferral list and told to refrain from donating blood, plasma, body organs, other tissue, or sperm. In addition, all donors with positive test results were contacted and advised to consult with their physician for further evaluation and treatment.

Testing showed that HTLV-III antibodies were present in the serum of donors whose blood had been given to patients with transfusion-associated AIDS, and that antibodies were almost always present in the serum of donors suspected to be a source of AIDS [31]. For example, investigation of a 60-year-old woman who developed AIDS after transfusion showed that the HTLV-III antibody was present in her blood as well as in the saliva and blood of the man who donated the blood [32]. Another study reported that three newborn infants received transfusions from an apparently healthy donor whose blood was subsequently found to be positive for the HTLV-III antibody. Eventually, all three infants became ill, with evidence of immunologic dysfunction, and HTLV-III antibodies were discovered in their blood [33]. In another study, the HTLV-III/LAV virus was isolated from the blood of 22 of 25 donors considered to be in the high-risk category for developing AIDS [34], supporting the use of the antibody test to screen potential donors and blood products for the risk of disease transmission.

Despite this evidence of the utility of the HTLV-III antibody test, blood bank directors were reluctant to use this test for donor screening. They were concerned that false-positive test results would be very traumatic for potential donors and inspire fear of the test in others, resulting in a decrease in the pool of voluntary donors [35]. Furthermore, if a donor did test positive, who would be responsible for providing counseling and follow-up of the donor and any sexual or other contacts? It was also recognized that at-risk persons might visit blood donation centers merely to discover their antibody status [36], which would be disruptive to collection procedures and dissuade others from donating. The impact on the blood supply was another consideration; removal of suspect units of blood and plasma might result in blood shortages and affect the ability of the blood bank to meet the requirements of the community. Because blood cannot be stored for more than a few weeks, a decrease of even a small number of donors could lead to critical shortages. Finally, it was argued that transfusion-associated AIDS had been reported by only a few states, so why should the recommendations of the US Public Health Service to implement the antibody screening test apply to all blood banks?

Although all of these arguments for not adopting HTLV-III antibody testing have some merit, they pale in comparison to the risk that untested blood might transmit a devastating infection to one or more recipients. For example, multiple blood transfusions were given to a boy with hemophilia at the time of his birth in January 1985 [37]. Routine testing of blood for antibodies to the virus did not begin until July 1985, just 6 months after his transfusions, but it was too late for him. He tested positive for the virus and developed full-blown AIDS and lymphatic cancer by the age of 7.

By the end of 1985, blood banks were screening all donated blood for HTLV-III antibody and discarding units with positive test results, and the FDA was able to report that infection of the blood supply was greatly diminished; only 1 in 400 units of blood was antibody-positive [38]. Furthermore, only 2831 of 1.1 million units of donated blood (0.25%) were repeatedly antibody-positive, and the majority of antibody-positive donors (89%) had identifiable risk factors for HTLV-III infection. The prevalence of antibody-positive samples in blood donors reported by Red Cross blood regions [39] is shown in the Table:

Red Cross blood centers	Blood samples with HTLV-III antibodies (%)
Los Angeles, Washington, DC	0.11
Boston, Detroit, Philadelphia	0.03
Portland, OR	0.015
Peoria, IL	0.011
Tulsa, OK	0.003

The percentage of positive samples correlates with the prevalence of infected individuals in the various geographical areas: the highest are in large urban areas and the lowest are in small rural communities.

During the next few years, the screening tests for HIV antibodies became more widespread. However, no test is perfect, and there were occasional false-positive and rarely false-negative results [40]. One of the hemophilia patients attending the Northwestern clinic initially had a positive test result but remained well and without any signs of infection over a 2-year period. He was retested, and the result was negative, and it was confirmed to be negative by repeated testing, suggesting that the original test result was a false-positive. Over time, the sensitivity and specificity of the tests for HIV improved, and it was shown that hemophiliacs with repeated negative results on antibody testing did not have HIV in their blood, even if they had a history of exposure to clotting factor concentrates [41].

Although it was clear that the virus was transmitted by sexual contact, by an infected woman to her unborn child, and by blood, reports that the virus had been isolated from saliva and tears triggered fears that kissing and even shaking hands could result in infection [42]. Despite abundant evidence that AIDS was not spread by casual contact (eg, no infections in health care workers except rarely after puncture by a bloody needle or spillage of infected blood onto unprotected skin or the mouth [43]), discrimination against hemophiliacs became widespread. Boys were excluded from classrooms (the Ryan White experience was common; eg, three brothers with hemophilia in Florida were barred from entering school), shunned by their friends, and subjected to social isolation. To counter this public response, the NHF assembled a scientific panel to advise it on the risks of AIDS in the school setting [44]. The panel concluded that it was virtually impossible to transmit or catch the disease from casual school contact. Also, the CDC stated that the benefits of an unrestricted educational setting outweighed the virtually nonexistent risk of transmission of the virus to other children or teachers, or the risk of the hemophilic child acquiring potentially harmful infections. They recommended against mandatory screening as a condition for school entry [45]. Confirmation that boys with hemophilia did not pose a risk to others came from the experience of a French boarding school between 1983 and 1985 [46]. Of the 44 children living at the school, infection was diagnosed in 12 of 24 hemophiliacs receiving regular treatment with clotting factor concentrates but in none of their 20 nonhemophilic roommates.

Although the HTLV-III antibody test could identify persons exposed to the virus, detection of the virus in the bloodstream was a better way to assess the potential for viral transmission. Research disclosed that some individuals with detectable virus had negative antibody test results. This was due to a "window" period; the time between infection and the development of antibodies. Studies suggested that this interval could range from 4 weeks to more than 6 months. These observations had major ramifications for efforts to prevent the spread of the disease. It meant that persons from high-risk groups, such as homosexuals and drug abusers, should not donate blood even if the antibody test was negative—they might be in the "window" period. This also applied to the wives of hemophiliacs; they were advised not to donate blood or become pregnant for fear that they might transmit the virus to their unborn infant [47,48].

Currently, the FDA attempts to exclude HIV-infected blood donors by requesting that they read the following and not donate if any of the following apply to them:

- Have AIDS or have ever had a positive HIV test result
- Have ever used needles to take drugs, steroids, or anything not prescribed by your doctor
- Male who has had sexual contact with another male, even once, since 1977[1]
- Have ever taken money, drugs, or other payment for sex since 1977
- Have had sexual contact in the past 12 months with anyone described in this list
- Have had syphilis or gonorrhea in the past 12 months
- In the past 12 months have been in juvenile detention, lockup, jail, or prison for more than 72 hours

---

1. On December 23, 2014, the FDA proposed that the indefinite donor deferral period for men who have sex with men should be reduced to 1 year since the last sexual contact.

- Have any of the following conditions that can be signs or symptoms of HIV/ AIDS:
  - Unexplained weight loss or night sweats
  - Blue or purple spots in the mouth or on the skin
  - Swollen lymph nodes for more than 1 month
  - White spots or unusual sores in the mouth
  - Cough that won't go away or shortness of breath
  - Diarrhea that won't go away
  - Fever of more than 100.5°F for more than 10 days

The donor information concludes with the following statement:

Remember that you CAN give HIV to someone else through blood transfusions even if you feel well and have a negative HIV test. This is because tests cannot detect infections for a period of time after a person is exposed to HIV. If you think you may be at risk for HIV/AIDS or want an HIV/AIDS test, please ask for information about other testing facilities. PLEASE DO NOT DONATE TO GET AN HIV TEST! [49].

Concern about transmission of HIV is heightened by the knowledge that 1 unit of plasma contaminated with HIV from an asymptomatic donor contains as many as 7500 infective doses of virus, and contains 875,000 if from a symptomatic donor [50]. The abundance of virus in the plasma explains the high frequency of infection in recipients of this material. However, the screening tests for HIV have been very effective; in 1996, it was estimated that the risk of giving blood during the infectious period for HIV was only 1 in 493,000 [51]. With further refinements in testing, the risk of transfusion-transmission of HIV has decreased to 1 in 2 million during the past decade [52].

KEY POINTS

- HIV-positive blood donors infected the national blood supply beginning in 1978. Sources of the virus included individual blood transfusions, fresh-frozen plasma, and plasma used for concentrate preparation.
- However, it was not until March 1983 that members of high-risk groups for AIDS were discouraged from donating blood, and most blood bankers resisted using laboratory tests to screen donors at that time.
- Concern about transfusion-transmitted infection led to consideration of "directed" donations, whereby a potential recipient would request blood only from designated donors. However, this provided a false sense of security about the safety of the product.
- Fear of AIDS contagion became widespread and persons with AIDS became social outcasts, despite evidence that the disease was not transmitted by casual contact.

- A major step forward was the introduction of the HIV antibody test in 1984. This test confirmed that transfusion of HIV-infected blood induced AIDS in recipients.
- It was not until July 1985 that routine HIV screening was instituted by blood collection agencies.

REFERENCES

[1] Centers for Disease Control. Possible transfusion-associated acquired immune deficiency syndrome (AIDS)-California. MMWR Morb Mortal Wkly Rep 1982;31:652–4.

[2] Ammann AJ, Wara DW, Dritz S, et al. Acquired immunodeficiency in an infant: possible transmission by means of blood products. Lancet 1983;i:956–8.

[3] Jett JR, Kuritsky JN, Katzmann JA, Homburger HA. Acquired immunodeficiency syndrome associated with blood-product transfusions. Ann Intern Med 1983;99:621–4.

[4] Marx JL. Health officials seek ways to halt AIDS. Science 1983;219:271–2.

[5] Greenwalt TJ. Blood-products transfusion and the acquired immunodeficiency syndrome. Ann Intern Med 1984;100:155.

[6] Iglehart JK. The centers for disease control. N Engl J Med 1983;308:604–8.

[7] Curran JW, Lawrence DN, Jaffe H, et al. Acquired immunodeficiency syndrome (AIDS) associated with transfusions. N Engl J Med 1984;310:69–75.

[8] Gordon Jr RS. More on blood transfusion and AIDS. N Engl J Med 1984;310:1742.

[9] Allan J. Quoted in the CCBC Newsletter; June 8/15, 1984.

[10] Gascon P, Zoumbos NC, Young NS. Immunologic abnormalities in patients receiving multiple blood transfusions. Ann Intern Med 1984;100:173–7.

[11] Deresinski SC, Cooney DP, Auerbach DM, Ammann AJ, Luft B, Goldman H. AIDS transmission via transfusion therapy. Lancet 1984;i:102.

[12] Duesberg P. Retroviruses as carcinogens and pathogens: expectations and reality. Cancer Res 1987;47:1199–220.

[13] Cohen J. Duesberg and critics agree: hemophilia is the best test. Science 1994;266:1645–6.

[14] Ward JW, Deppe EA, Samson S, et al. Risk of human immunodeficiency virus infection from blood donors who later developed the acquired immunodeficiency syndrome. Ann Intern Med 1987;106:61–2.

[15] Goedert JJ, Kessler CM, Aledort LM, et al. A prospective study of human immunodeficiency virus type I infection and the development of AIDS in subjects with hemophilia. N Engl J Med 1989;321:1141–8.

[16] Chigwedere P, Seage III GR, Gruskin S, Lee TH, Essex M. Estimating the lost benefits of antiretroviral drug use in South Africa. J Acquir Immune Defic Syndr 2008;49:410–15.

[17] South Florida Blood Service. Phone line guards blood supply and donor privacy; Lifeline/Winter 1984.

[18] Shilts R. And the band played on. New York, NY: St Martin's Press; 1987, p. 308.

[19] Council of Community Blood Banks. Cutter Laboratories, San Francisco area blood centers start anti-HBc testing. CCBC Newsletter; April 6, 1984.

[20] American Red Cross, American Association of Blood Banks, Council of Community Blood Centers, Joint News Release; June 22, 1983.

[21] Counts RB, Giblett ER. Designated donations. CCBC Newsletter; June 8/1/5, 1984.

[22] Bove JR. Transfusion-associated AIDS—a cause for concern. N Engl J Med 1984;310:115–16.

[23a]   Kruskall MS, Churchill WH, Brauer M, et al. Designated blood donations. N Engl J Med 1984;310:1194–5.

[23b]   Dahlke MB. Designated blood donations. N Engl J Med 1984;310:1195.

[24]   Starkey JM, MacPherson JL, Bolgiano DC, Simon ER, Zuck TF, Sayers MH. Markers of trans-mission-transmitted disease in different groups of blood donors. JAMA 1989;269:3452–4.

[25]   Retrieved from: <http://en.wikipedia.org/w/index.php?title=Ryan_White&aldid=577620834>; October 2013.

[26]   Friedland GH, Saltzman BR, Rogers MF, et al. Lack of transmission of HTLV-III/LAV infection to household contacts of patients with AIDS or AIDS-related complex with oral candidiasis. N Engl J Med 1986;314:344–9.

[27]   Health Resources and Services Administration (HRSA), HIV/AIDS Bureau (HAB), US Department of Health and Human Services <www.hhs.gov>.

[28]   CDC, FDA, ADAMHA, NIH, HRSA. Provisional Public Health Service inter-agency rec-ommendations for screening donated blood and plasma for antibody to the virus causing acquired immunodeficiency syndrome. MMWR Morb Mortal Wkly Rep 1985;34:5–7.

[29]   Licensing of the antibody test for HIV was expected in early February 1985, but was delayed until March because manufacturers of the test needed to provide additional information for the FDA to review. Reported by Altman LK, NY Times; February 15, 1985.

[30]   Goedert JJ. Blood donation by persons at high risk of AIDS. N Engl J Med 1985;312:1190.

[31]   Jaffe HW, Francis DP, McLane MF, et al. Transfusion-associated AIDS: serologic evidence of human T-cell leukemia versus infection of donors. Science 1984;223:1309–12.

[32]   Groopman JE, Salahuddin SZ, Sarngadharan MG, et al. Virologic studies in a case of transfusion-associated AIDS. N Engl J Med 1984;311:1419–22.

[33]   Wykoff RF, Pearl ER, Saulsbury FT. Immunologic dysfunction in infants infected through transfusion with HTLV-III. N Engl J Med 1985;312:294–6.

[34]   Feorino PM, Jaffe HW, Palmer E, et al. Transfusion-associated acquired immunodeficiency syndrome. N Engl J Med 1985;312:1293–6.

[35]   Osterholm MT, Bowman RJ, Chopek MW, McCullough JJ, Korlath JA, Polesky HF. Screening donated blood and plasma for HTLV-III antibody. N Engl J Med 1985;312:1185–9.

[36]   Anonymous. HTLV-III and blood donors. Lancet 1985;i:856.

[37]   Wolinsky H. Boy fights for normal life despite diseases. Chicago Sun-Times July 17, 1994.

[38]   Anonymous. Progress on AIDS. FDA Drug Bull 1985;15:27–32.

[39]   Schorr JB, Berkowitz A, Cumming PD, Katz AJ, Sandler SG. Prevalence of HTLV-III anti-body in American blood donors. N Engl J Med 1985;313:384–5.

[40]   Taylor H-L, Moulsdale HJ, Mortimer PP. Blood donor screening by Wellcome anti-HIV kits. Lancet 1987;i:631–2.

[41]   Gibbons J, Cory JM, Hewlett IK, Epstein JS, Eyster ME. Silent infections with human immunodeficiency virus type 1 are highly unlikely in multi-transfused seronegative hemo-philiacs. Blood 1990;76:1924–6.

[42]   Sande MA. Transmission of AIDS: the case against casual contagion. N Engl J Med 1986;314:380–2.

[43]   Anonymous. Update: human immunodeficiency virus infections in health-care workers exposed to blood of infected patients. MMWR Morb Mortal Wkly Rep 1987;36:285–9.

[44]   Hemophilia Information Exchange. The National Hemophilia Foundation's panel concludes that school attendance by children with AIDS presents no risks; November 18, 1985.

[45]   Anonymous. Education and foster care of children infected with human T-lymphotrophic virus type III/lymphadenopathy-associated virus. MMWR Morb Mortal Wkly Rep 1985;34:517–21.

[46] Berthier A, Fauchet R, Genetet N, et al. Transmissibility of human immunodeficiency virus in haemophilic and non-haemophilic children living in a private school in France. Lancet 1986;ii:598–601.

[47] Hemophilia Information Exchange. HTLV-III virus and sexual partners of persons from high risk groups; May 17, 1985.

[48] Ragni MV, Rinaldo CR, Kingsley LA, et al. Heterosexual partners of haemophiliacs must refrain from blood donation. Lancet 1986;i:1033.

[49] FDA. Blood donor educational materials: MAKING YOUR BLOOD DONATION SAFE; December 2012.

[50] Ho DD, Moudgil T, Alam M. Quantitation of human immunodeficiency virus type I in the blood of infected persons. N Engl J Med 1989;321:1621–5.

[51] Schreiber GB, Busch MP, Kleinman SH, Korelitz JJ. The risk of transfusion-transmitted viral infections. N Engl J Med 1996;334:1685–90.

[52] Goodnough LT, Shander A, Brecher ME. Transfusion medicine: looking to the future. Lancet 2003;361:161–9.

Chapter 7

A Full-Blown Epidemic

By December 1983, there were a total of 21 hemophilia patients with confirmed AIDS and 30 highly suspect transfusion cases. Fig. 7.1 shows that the number of confirmed diagnoses of AIDS in hemophiliacs increased from 1981 through 1984, almost doubling every 6 months [1].

The low death rate prior to 1984 misled some people into thinking that only a few hemophiliacs had contracted the infection. At the time, the relatively long incubation period between exposure to the virus and immune depletion was not appreciated; hemophiliacs died of AIDS only when immunodeficiency had become so severe that fatal infections occurred, and this could take years. Because there was no blood test to show if the virus was present in an individual, it was impossible to know how many hemophiliacs were actually infected. That infection was indeed quite prevalent was shown by the sharp increase in the number of cases (42) and the number of deaths (22) recorded in June 1984.

Although there was no direct evidence that clotting factor concentrates transmitted AIDS, there were several reasons for suspecting that exposure to blood products was responsible. First, the distribution of AIDS cases was similar to the pattern of hepatitis B virus infection, which was known to be transmitted by blood and needles contaminated by blood [2]. Second, AIDS occurred in hemophiliacs whose only risk factor appeared to be the use of factor VIII or factor IX concentrates [3,4]. Furthermore, studies of hemophiliacs receiving concentrates showed that they had alterations in the distribution of their immune cells that were similar to the changes found in nonhemophiliacs with AIDS risk factors [5a–d]. As the weight of evidence implicating blood products in the transmission of AIDS accumulated, blood banks and pharmaceutical companies developed procedures to identify and exclude high-risk donors, including homosexuals, Haitians and recent visitors to Haiti, and users of illicit intravenous drugs. One manufacturer, Armour Pharmaceutical Company, reported that the plasma used for production of their products was not obtained from donor sites located in "high-risk" urban areas, such as New York, San Francisco, Los Angeles, and Miami. Further, they did not collect plasma from prisoners and used only original source material [6]. Alpha Therapeutic Corporation likewise required all its plasma suppliers to screen donors for AIDS risk factors and AIDS, and it denied using plasma collected from prisoners [7]. Despite these

Linked by Blood: Hemophilia and AIDS. DOI: http://dx.doi.org/10.1016/B978-0-12-805302-7.00007-0

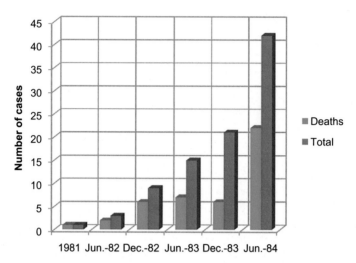

FIGURE 7.1   AIDS in hemophiliacs, 1981–84. *Data from the Morbidity and Mortality Weekly Report, June 26, 1984.*

precautions, the number of US hemophiliacs with AIDS continued to increase and reached 21 by December 1983 [8].

The commercially prepared concentrates manufactured in the United States were sold in European countries. If concentrates were responsible for transmitting AIDS to US hemophiliacs, then they might also infect hemophiliacs in other countries. I conducted a survey of European Hemophilia Centers in 1983 [9]. Five cases of possible AIDS were identified, as were five patients with immune thrombocytopenia, a condition that had recently been described in hemophiliacs treated with clotting factor concentrates and possibly due to altered immunity [10]. By December 1983, fatal AIDS cases in two patients treated with concentrate were reported in England and Japan [11a,b]. In addition, alterations in immune function were described in 11 of 16 British patients receiving commercial blood products of American origin [12]. Although the number of affected patients was small compared to the total number of hemophiliacs, David L. Aronson of the US Office of Biologics observed that there had been an increase in the number of pneumonia deaths in US and British hemophiliacs since 1975, and he speculated that some of these could have been associated with AIDS [13].

However, as of June 1984, only 11 of the 13,000 (0.08%) hemophiliacs living in Europe were diagnosed with AIDS, and because of this small number it was felt that concentrate use should not be abandoned [14]. This was because of the constant risk of fatal bleeding in severe hemophiliacs, which was as high as 0.9% during this time [15,16]. The expert opinion was that a "hypothetical danger" should not overrule the "immense benefit" of concentrates. British investigators questioned whether the immunologic abnormalities in hemophiliacs were due to an infectious agent and posited that the immune system was simply reacting to proteins present in the clotting factor concentrates [17]. This view was supported by an editorial in a leading medical journal, *The Lancet*. The editorialists wrote that "the recognition of disease (AIDS) in a few hemophiliacs does not necessarily reflect the tip of an iceberg," and they supported the position of the US National Hemophilia Foundation (NHF) that there should be no change in treatment policy [18]. However, studies published in the latter part of 1984 reported that approximately 60–70% of hemophiliacs in Europe and the United States had evidence of exposure to the AIDS virus, and all had been treated with factor VIII concentrates [19a,b].

In the autumn of 1983, the Medical and Scientific Advisory Council (MASAC) of the NHF distributed recommendations to prevent AIDS in patients with hemophilia [20]. They supported the use of cryoprecipitate in newborns and children younger than 4, in patients not previously exposed to factor VIII concentrate, and in those with less severe hemophilia treated infrequently. They noted that the choice of cryoprecipitate versus concentrate for older children and men with severe hemophilia was controversial and made no specific recommendations for these individuals. They urged concentrate makers to exclude donors from groups at risk for AIDS, and they advised direct questioning and physical examination of donors to detect symptoms associated with AIDS. They encouraged the development of tests to identify high-risk donors and recommended that manufacturers should avoid collecting plasma at donor centers located in geographic areas with a high incidence of AIDS. In addition, they offered suggestions about reducing the size of the donor pool used for preparing concentrates. Finally, they appealed to blood centers located in regions with a low incidence of AIDS to increase their capacity for cryoprecipitate production; this material could be used locally and shared with other regions.

In September 1984, the NHF prepared an informational pamphlet entitled *AIDS and Hemophilia: Your Questions Answered* [21]. The disease, the virus, and the risks of getting AIDS were described, as well as a set of actions to be taken in response to the disease. In retrospect, it seems that the risks of contracting AIDS from concentrate were downplayed. For example, it is stated on the first page that "some data … suggest that it [AIDS] may be transmitted from infected individuals through blood and blood products, and intimate contact with certain body secretions. However, many people appear not to be susceptible to AIDS, or to develop such mild symptoms that no disease is detectable after exposure to the virus." This statement was based on the observation that although many hemophiliacs tested positive for antibodies to HIV, only a few had developed AIDS. But NHF should have cautioned that it would be impossible to know whether someone was susceptible to developing AIDS unless they were observed for at least 5 years. That is the length of the incubation period prior to the appearance of AIDS that had been established by the CDC in December 1983, almost a year before the pamphlet was published. In response to the question, "Can one do anything to minimize the risk of AIDS?" the pamphlet states that "we cannot recommend any specific change in treatment or in life-styles that will definitely lessen the risk of developing this syndrome." In particular, the pamphlet urged that hemophiliacs should not change or stop treatment with clotting factor and stated that there was no specific evidence that there was a greater risk with concentrate than with cryoprecipitate or fresh-frozen plasma. This view was contested by Dr Oscar Ratnoff, who reported that in his patients, AIDS was much more frequent with concentrate than with cryoprecipitate treatment [22]. However, it was not until September 1985 that a study supporting his observations was published. It revealed that antibodies to the virus were detected in 40% of those treated solely with cryoprecipitate as compared to 77% of those receiving concentrate [23]. This study confirmed that both cryoprecipitate and concentrate could transmit the virus, but that the risks were greater with concentrate. However, people were reluctant to switch to cryoprecipitate because it was not always available from local blood banks and it was tedious to thaw and infuse.

There were other physicians concerned about the continued use of clotting factor concentrates. Dr Jane Desforges editorialized in the *New England Journal of Medicine* that preventing the complications associated with concentrates might have to take precedence over preventing the complications of hemophilia, and she suggested that home treatment, with its strong dependence on the use of commercial concentrates, should be abandoned [24]. However, most hemophiliacs and their caregivers were very reluctant to surrender the treatment that had transformed their lives and the lives of their families. Furthermore, people with factor IX deficiency (approximately 20% of all hemophiliacs) who wished to treat at home almost always had to use commercial concentrates. Their only other choice was fresh-frozen plasma, which was cumbersome to use because it had to be infused in large amounts to control bleeding; cryoprecipitate was not an option for these individuals because it does not contain factor IX.

In October 1984, NHF issued a Chapter Advisory containing a notification from Hyland Laboratories and the Red Cross that that they were recalling several lots of concentrate because the plasma used in its production came from a donor who had died of AIDS [25]. There was the possibility that some of these lots might have already been used for treatment of hemorrhages. This announcement further eroded confidence in the safety of commercial clotting factor concentrates. Although the NHF might have used this opportunity to caution their members about the potential hazards of concentrates, they instead noted that the product recall actions were taken as a strictly precautionary measure and that patients should continue to use concentrate or cryoprecipitate.

Recognizing that concentrates posed a risk of transmission of hepatitis, if not AIDS, pharmaceutical companies investigated whether heating the material would reduce its infectivity. Researchers had determined that factor IX preparations could be heated without loss of activity, but factor VIII was thought to be more vulnerable. However, as early as 1983, Hyland Laboratories succeeded in developing a heat-treated factor VIII product that provided partial protection against hepatitis. Because there was no test for HIV at the time, it could not be determined if this product transmitted the AIDS virus. Nevertheless, the FDA licensed this heat-treated concentrate in August 1983, but it was available in only limited supply. Edmond S. Smith, the Chief of California Children Services, recommended that priority for the use of heated concentrate should be given to newly diagnosed patients, those requiring infrequent treatment, patients needing surgery, and those with AIDS-like symptoms [26].

However, NHF's MASAC resisted recommending the heated concentrate; members felt that there was insufficient efficacy data to support the replacement of standard factor VIII concentrates by heat-treated products [27]. Their past surveys showed that uncontrolled bleeding was the major cause of death for hemophiliacs, and AIDS was a new phenomenon of uncertain lethality. In addition, they noted that the price of the heated product was twice that of nonheated material [28]. Furthermore, there was concern that heating might modify the clotting factor, thus altering the patient's response to treatment. They used these arguments to support their recommendations for the continued use of the currently available clotting factor concentrates. However, they did encourage the development of methods for rendering blood products free of infectious agents, and they urged that prospective clinical trials should be conducted to establish the safety and efficacy of heat-treated products. A survey of MASAC members conducted in May 1984 found unanimous support for not changing their position on heat-treated products [29].

The isolation of the HIV virus in 1983 was a major step forward in the manufacture of safer clotting factor concentrates. It enabled the selection and testing of techniques that would spare the clotting factor but destroy the virus, and it led to the discovery that pasteurization of freeze-dried concentrates would accomplish this objective. The procedure is similar to the method used to decrease the number of viable bacteria in milk; it involves heating to a specific temperature for a predetermined length of time. In October 1984, Cutter Biologicals announced that factor VIII concentrate containing AIDS-related viruses could be sterilized by heating for 72 hours at 68°C (154°F), with preservation of most of the original clotting activity [30]. They added virus to factor VIII concentrate and demonstrated that after heating, the virus or its active enzyme became undetectable. Cutter immediately began to heat-treat all their concentrates, and other manufacturers soon followed suit [31]. The CDC, FDA, and independent investigators confirmed that prolonged exposure to high temperatures killed the virus [32,33]. The effectiveness of heat treatment was confirmed by an observational study of outcomes in two groups of previously untreated patients; one group had received a variety of unheated concentrates and the other was treated exclusively with heated material. Five of twenty-nine in the first group and none of eighteen in the second group developed lymphadenopathy associated virus (HIV) antibodies [34].

Heat-treated concentrates became generally available in the autumn of 1984. However, there was concern that exposure of infants and young children to a concentrate, even though heated, might increase the risk of hepatitis; in fact, signs of hepatitis were reported in 11 of 13 hemophiliacs treated exclusively with heated concentrate [35]. In addition, there was evidence that heat treatment might damage the fragile clotting protein [36], potentially inducing antibodies to the altered factor and the development of resistance to further treatment. These considerations led some workers to advocate for the continued use of unheated material [37]. However, it was recognized that all unheated concentrates posed an infection risk not only to patients but also to those preparing and infusing them [38]. The Medical Advisory Council of the Illinois Hemophilia Foundation met on October 16, 1984, and concluded that heated concentrates were safer than nonheated concentrates with regard to transmission of AIDS but not hepatitis [39]. Furthermore, the heated concentrates were as effective in controlling bleeding and did not appear to increase the risk of developing treatment resistance. The Illinois Hemophilia Foundation provisionally recommended the use of heat-treated concentrates as preferable to nonheated concentrates.

The NHF's MASAC met in October 1984, and suggested that because there were no untoward effects attributable to heated concentrates, hemophilia physicians should strongly consider changing to these products with the understanding that protection against AIDS was yet to be proven [40]. But at a subsequent meeting in February 1985, they did not endorse the use of heated concentrates in all hemophiliacs; instead, they limited their recommendation to young children and newly identified adults with severe hemophilia [41,42]. They specified that heated concentrates should replace cryoprecipitate or unheated concentrates because:

- Infants exposed to HTLV-III might be at increased risk for developing AIDS and therefore should receive products least likely to be contaminated with the virus (ie, heated concentrate).
- Cryoprecipitate should be avoided because it had been shown to transmit HTLV-III and AIDS.
- Heat treatment appeared to be very effective at killing HTLV-III.

Despite these apparent advantages of heat-treated concentrates, MASAC continued to urge adult hemophiliacs to infuse unheated concentrate or cryoprecipitate as prescribed by their physicians. At the meeting, there was considerable discussion about whether heated concentrates should replace unheated concentrates for all individuals with hemophilia. This switch to heat-treated concentrates had been championed by a recent editorial in *The Lancet* [43]. But one expert noted that larger amounts of heated than nonheated concentrate were required to achieve therapeutic levels of the clotting factor, and these higher doses would increase treatment costs. There was also concern about the limited availability of the heated concentrates. The members were evenly split on the issue; because there was no consensus, no action was taken. In the absence of definitive recommendations, unheated lots of concentrate, some prepared from donors with confirmed AIDS [44], continued to be infused and were not completely withdrawn until June 1985 [45]. It would be another 4 years before there was definitive evidence that heat-pasteurized concentrates did not transmit HIV infection [46].

Finally, in April 1985, MASAC recommended that heat-treated concentrate should be prescribed for every individual with severe hemophilia [47]. The heat-treated concentrates would replace cryoprecipitate as well as unheated concentrates. This action was prompted by a new report that 40% of hemophiliacs treated exclusively with cryoprecipitate had antibodies to HTLV-III [48]. In addition to the new evidence that cryoprecipitate could transmit AIDS, NHF had two other reasons for making this policy change: (1) the AIDS virus appeared to be adequately killed by the newly-licensed heat treatment and (2) the withdrawal of several lots of unheated factor VIII and factor IX concentrates due to confirmed AIDS in a donor had further undermined confidence in the safety of unheated material [49].

NHF updated its "Your Questions Answered" early in 1985, but it continued to downplay the risks of infection. For example, to answer the question, "How does one get AIDS?" they wrote: "The great majority of people who are exposed to the AIDS agent appear not to be susceptible to AIDS...." They probably based this assessment on the fact that individuals could have antibodies to the virus but not have symptoms of illness. However, as previously noted, there is a long latent period between the time of infection and the development of AIDS; this information had been available for more than a year and should have prompted NHF to be more cautious in their reassurances to members about the risks of getting AIDS. A more egregious error was NHF's response to "Does the finding of antibodies to HTLV-III indicate a person has AIDS?" by stating that "Having antibodies simply indicates that one has been exposed to the virus and had an immune response to it. It does not necessarily mean that one is infected..." This was incorrect; a study published in November 1984 concluded that the antibody test recognized a virus that could be carried and transmitted and was an important cause of AIDS [50]. Antibody-positive persons should not have been told that they were not infected; this might have led to potentially dangerous practices such as unprotected sex and blood donation, and it provided a false sense of security. A few months later, a brochure distributed by the Public Health Service answered the question of "Is there a laboratory test for AIDS?" by stating unequivocally: "Presence of HTLV-III antibodies means that a person has been infected with the AIDS virus" [51].

During the latter half of 1984, the number of hemophiliacs with AIDS progressively increased [52]. There were 58 confirmed cases and 31 deaths; the incidence rate was 3.6/1000 patients. By 1985, it was suspected that more than 1 million persons in the general population had been exposed to the AIDS virus, and this included a majority of those with hemophilia. In fact, hemophiliacs appeared more vulnerable than homosexuals or drug abusers [53]; it was estimated that as many as 70% of all hemophiliacs were infected [54]. There were also reports that the virus could be transmitted to the family members of an infected hemophiliac; the 71-year-old wife of a man dying of AIDS developed recurrent infections typical of the immunodeficiency syndrome [55]. On further investigation, 10% of the wives of hemophiliacs tested positive for HTLV-III [56]. In addition, AIDS was reported in the 5-month-old son of a hemophiliac, and both parents had evidence of pre-AIDS [57]. In response to these reports, the CDC recommended that the wives of hemophiliacs should be tested for HTLV-III antibodies [58]. Women with positive test results were advised to defer pregnancy, and those with negative outcomes were told to avoid unprotected sex. Two years later, the CDC reported that 77 of 2276 (3%) spouses tested positive. Further investigation showed that 22 became pregnant and had 20 children; 9 of these children were infected with the virus [59]. However, testing was negative for children of infected men whose wives were not infected [60], confirming that the virus was not spread by casual contract.

In May 1985, Dr M. Elaine Eyster and NHF distributed an AIDS Update to all providers of hemophilia care; it communicated information about HIV transmission to the spouses of hemophiliacs [61]. In this Medical Bulletin, Dr Eyster recommended that hemophiliacs should use condoms and postpone pregnancies. Because NHF leadership was concerned that this material would be upsetting for many hemophiliacs and their sexual partners, they suggested that it should be presented by staff at chapter meetings rather than at one-on-one sessions with members. The inadequacy of this indirect method of communication was revealed when a study performed 2 years later found that only 40% of couples used condoms consistently and pregnancies occurred in 9.5% [62]. Subsequently, the CDC recommended that hemophilic patients and their sexual partners should receive thorough, confidential, and individualized counseling [63], a position endorsed by the NHF [64]. Evidence that sexual education and counseling programs reduced exposure to the virus was provided by the subsequent observation that none of 87 household contacts of 68 HIV antibody–positive hemophiliacs had acquired the infection [65].

All hemophiliacs with positive antibody test results were also found to have detectable HIV in their blood [66], and studies showed progressive deterioration of their immune cell function [67,68]. Furthermore, skin tests revealed that half of all severe hemophiliacs lacked immunity to common bacterial proteins [69]. Those with impaired immunity had used more clotting factor concentrate and were more likely to have been exposed to HTLV-III [70]. Most patients developed AIDS within 2–5 years of becoming infected by the virus [71,72]. Older age at the time of infection was more likely to be associated with more severe immunologic defects and a more rapid onset of AIDS [73]. In a group of 28 hemophiliacs followed for up to 9 years, CD4 cells declined at a rate of 13.5% per year, and this decline was significantly faster in those 25 years of age or older as compared to those younger than 25 (17.5% per year vs 9.5% per year). The purity of the clotting factor concentrate also affected the rate of decline in immune function. Some clotting factor concentrates contained a variety of extraneous proteins in addition to the clotting factor and were designated as being of "intermediate purity," whereas others were more highly refined and termed "high purity." Intermediate-purity concentrates were associated with a more rapid decrease in immune cells, perhaps because the higher protein content had a more suppressive effect on the immune system [74,75].

Persons with hemophilia and their families questioned the safety of concentrates and were concerned that blood banks and pharmaceutical companies were not exerting every effort to provide safe products. Some of those contracting AIDS instituted lawsuits against physicians and concentrate manufacturers, claiming failure to warn of potential risks [76]. Physicians defended themselves by arguing that even if they had known of the risks, they still would have prescribed the blood products because of the dangers of uncontrolled bleeding. Pharmaceutical companies, however, were held to a different standard; they have a continuing duty to warn of product hazards. The Canadian Commission of Inquiry on the Blood System ruled that physicians and the general population had not been adequately informed about AIDS and hepatitis risks by the manufacturers, who eventually made a large financial settlement [77].

The NHF encouraged physicians to be conscientious in reporting new cases to the CDC, because AIDS still was not an officially reportable disease. However, NHF indicated that physicians should keep the information about the numbers of infected persons confidential because it did not want to alarm its members. This advice, that doctors should withhold certain information from their patients, was inconsistent with best medical practice, and the NHF would later be criticized for adopting a paternalistic attitude toward its membership (see chapter 8: The Institute of Medicine Study).

The lag period between infection with the virus and the appearance of AIDS continued to mislead epidemiologists. For example, the CDC suggested that the incidence of hemophilia-associated AIDS was stabilizing or declining, because only six new cases had been reported as of April 1985 [78]. But by May 1985, the total number of infected hemophiliacs had climbed to 73, and infection associated with the use of cryoprecipitate, as well as concentrate, was confirmed [79,80]. The chances of becoming infected increased with increasing exposure to concentrate [81,82]. The validity of this observation was confirmed by a European study reporting that heavily treated French hemophiliacs were significantly more likely to test positive for antibody than lightly treated Belgian hemophiliacs (63.8% vs 3.4%) [83]. Further examination showed that the highest rate of infection occurred in hemophiliacs treated with factor VIII concentrates prepared from American source plasma, intermediate values were found for those given European concentrates, and the lowest rates of infection were found for persons receiving red blood cells without plasma [84].

A retrospective study of hemophiliacs from western Pennsylvania found that 81.8% of those treated with factor VIII concentrate were HIV-antibody-positive compared to 48.1% treated with factor IX concentrates, 10% treated with cryoprecipitate, and none treated with fresh-frozen plasma [85]. Fig. 7.2 shows when 77 hemophiliacs in western Pennsylvania became exposed to the virus. No patients were infected in 1977, 28 were infected in 1982 (the peak

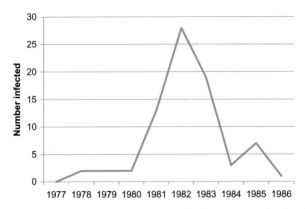

FIGURE 7.2   When 77 hemophiliacs in western Pennsylvania were exposed to the virus.

incidence), and only one was infected in 1986. Infections occurring between 1984 and 1986 were probably due to the continued use of older, contaminated plasma processed into cryoprecipitate or unheated concentrate.

The number of AIDS cases in Europe also began to accelerate, increasing at the rate of 10 cases per week, and it more than doubled in the interval from 1983 to 84 [86]. Seventeen of these cases were in hemophiliacs, and half of those were in Germany; hemophilia centers in Germany were major users of concentrate manufactured in the United States [87]. In the United Kingdom, 32% of hemophiliacs were found to be antibody-positive [86].

The widespread adoption of heat-treated concentrates that began in June 1985 effectively ended the transmission of new HIV infections, except in rare instances [88]. The impact of HIV on the longevity of persons with hemophilia was examined in 1994 by workers from the CDC [89]. They showed that the death rate increased from 0.5 per million in 1981 to 1.3 per million in 1989, and the median age at death decreased from 57 years in 1979–81 to 40 years in 1987–89. Deaths in hemophiliacs were no longer mainly due to bleeding; they were now attributed to infections in immune-compromised persons harboring HIV.

Another major advance was the introduction of recombinant human factor VIII for the treatment of hemophilia, which was first described in 1989 [90,91]. These two measures—the heating of concentrates and the use of recombinant factors—effectively ended the epidemic. The 2013 report of The American Thrombosis and Hemostasis Network (ATHN) noted there were no HIV infections in more than 6000 hemophiliacs born after 1985. However, more than 60% of those exposed to clotting factors in the form of cryoprecipitate or concentrate prior to 1985 became infected by the virus [92]. In 2014, nearly 20% of hemophilic men between the ages of 30 and 74 were living with HIV [93].

Although the majority of persons with hemophilia treated with clotting factor concentrates between 1979 and 1984 became infected with HIV, a small number remained virus-free, even when examined by the most sensitive assays [94]. A study of 560 such individuals failed to reveal a common genetic variant that could explain their resistance to HIV infection; only 6.4% had a genetic mutation linked to viral resistance in other populations [95]. How such individuals are able to remain free of infection despite exposure to the virus is being intensively investigated.

The introduction of azidothymidine (Zidovudine) in 1985 prevented the progressive deterioration of the immune system in HIV-infected hemophiliacs and dramatically decreased fatalities [96]. In the years that followed, more effective antiviral medications were discovered. In the early 1990s, the introduction of multi-drug regimens dramatically decreased the number of HIV-positive hemophiliacs developing AIDS. Research to enhance the survival of CD4+ T cells infected by HIV is currently in progress [97].

KEY POINTS

- The number of hemophiliacs with AIDS increased slowly between the end of 1981 and December 1983, but it increased sharply in June 1984. This was because most hemophiliacs were probably not exposed to the virus until 1978 at the earliest, and the long interval (average, 5.5 years) between infection and immune depletion resulted in the sudden appearance of AIDS cases in 1984.
- Clotting factor concentrates were suspected of transmitting AIDS because the distribution of AIDS cases was similar to transfusion-transmitted hepatitis. In addition, concentrate use seemed to be the sole risk factor for infection, and AIDS appeared in European hemophiliacs exposed solely to US concentrates.
- The risk of infection from blood or cryoprecipitate was small, because initially there were only a few HIV-infected donors. However, the risk from commercially prepared concentrates was high, because thousands of liters of plasma were required for plasma fractionation, and a single donation containing the virus could contaminate the entire lot. At least 77% of hemophiliacs infusing concentrate became infected by the virus.
- As late as December 1984, the NHF reiterated its position that the use of concentrates should be continued to preserve life and health.
- Factors that increased the risk of AIDS included older age, greater concentrate use, and products that were of intermediate purity rather than high purity.
- Some infected hemophiliacs transmitted the virus to their spouses, who passed the virus to their infants during subsequent pregnancies.
- The identification and characterization of the AIDS virus in 1983 were major steps forward in the development of safe clotting factor concentrates. Heat-treated concentrates were licensed in late 1984, and their use was

recommended by NHF in the spring of 1985. This ended the transmission of new HIV infections.

● By May 1985, the number of US hemophiliacs with AIDS had climbed to 73, and infection by exposure to cryoprecipitate, as well as concentrate, was confirmed. Antibodies to the virus were found in 82% of those exposed to concentrate but only in 10% of those treated solely with cryoprecipitate. By July 1987, the number with AIDS had increased to 374, and hundreds of others were asymptomatic but infected by the virus.

● Data collected in 1994 showed that the death rate in hemophiliacs increased from 0.5 per million in 1981 to 1.3 per million in 1989 because of HIV infection. The introduction of combinations of anti-HIV drugs in the 1990s led to a marked decline in fatalities.

REFERENCES

[1]   Fauci AS, et al. NIH conference. Acquired immunodeficiency syndrome: epidemiologic, clinical, immunologic, and therapeutic considerations. Ann Intern Med 1984;100:92–106.

[2]   Prevention of Acquired Immune Deficiency Syndrome (AIDS): Report of Inter-Agency Recommendations. MMWR Morb Mortal Wkly Rep March 4, 1983;32:101–3.

[3]   Polesky HF, Medical Director. War Memorial Blood Bank. Minneapolis, MN.

[4]   Goldsmith JM, Variakojis D, Phair JP, Green D. The spectrum of human immunodeficiency virus infection in patients with factor IX deficiency (Christmas disease). Am J Hematol 1987;25:203–10.

[5a] Ragni MV, Spero JA, Lewis JH, Bontempo FA. Acquired-immunodeficiency-like syndrome in two haemophiliacs. Lancet 1983;i:213–14.

[5b] Lederman MM, Ratnoff OD, Scillian JJ, Jones PK, Schacter B. Impaired cell-mediated immunity in patients with classic hemophilia. N Engl J Med 1983;308:79–83.

[5c] Menitove JE, Aster RH, Casper JT, et al. T-lymphocyte subpopulations in patients with classic hemophilia treated with cryoprecipitate and lyophilized concentrates. N Engl J Med 1983;308:83–6.

[5d] Goldsmith JC, Moseley PL, Monick M, Brady M, Hunninghake GW. T-lymphocyte subpopulation abnormalities in apparently healthy patients with hemophilia. Ann Intern Med 1983;98:294–6.

[6]   AIDS Policy Position, Armour Pharmaceutical Company.

[7]   Alpha Hemophilia Newsletter Winter 1982;4:1.

[8]   Centers for Disease Control (CDC) Update: acquired immunodeficiency syndrome (AIDS) among patients with hemophilia—United States. MMWR Morb Mortal Wkly Rep December 2, 1983;32:613–15.

[9]   Green D. AIDS and immunologic abnormalities in European and American Hemophiliacs. Scand J Haematol 1984;33(Suppl. 40):367–9.

[10] Ratnoff OD, Menitove JE, Aster RH, Lederman MM. Coincident classic hemophilia and "idiopathic" thrombocytopenic purpura in patients under treatment with concentrates of anti-hemophilic factor (factor VIII). N Engl J Med 1983;308:439–42.

[11a] Daly HM, Scott GL. Fatal AIDS in a hemophiliac in the UK. Lancet 1983;ii:1190.

[11b] Shibuya A, Saitoh K, Tsuneyoshi H, Tanaka S. Burkitt's lymphoma in a haemophiliac. Lancet 1983;ii:1432.

[12] Jones P, Proctor S, Dickinson A, George S. Altered immunology in haemophilia. Lancet 1983;i:120–1.

[13] Aronson DL. Pneumonia deaths in haemophiliacs. Lancet 1983;ii:1023.

[14] Bloom AL. Acquired immunodeficiency syndrome and other possible immunological disorders in European haemophiliacs. Lancet 1984;i:1452–5.

[15] Triemstra M, Rosendaal FR, Smit C, Van der Ploeg H, Briet E. Mortality in patients with hemophilia. Ann Intern Med 1995;123:823–7.

[16] Darby SC, Kan SW, Spooner RJ, et al. Mortality rates, life expectancy, and causes of death in people with hemophilia A or B in the United Kingdom who were not infected with HIV. Blood 2007;110:815–25.

[17] Carr R, Edmond E, Prescott RJ, et al. Abnormalities of circulating lymphocyte subsets in haemophiliacs in an AIDS-free population. Lancet 1984;i:1431–3.

[18] Anonymous. Acquired immunodeficiency in haemophilia. Lancet April 2, 1983;i:745.

[19a] Melbye M, Biggar RJ, Chermann JC, Montagnier L, Stenbjerg S, Ebbesen P. High prevalence of lymphadenopathy virus (LAV) in European hemophiliacs. Lancet 1984;ii:40–1.

[19b] Ramsey RB, Palmer EL, McDougal JS, et al. Antibody to lymphadenopathy-associated virus in haemophiliacs with and without AIDS. Lancet 1984;ii:397–8.

[20] National Hemophilia Foundation. Recommendations to prevent AIDS in patients with hemophilia. Hemophilia Information Exchange, Medical Bulletin #9, Chapter Advisory #12; revised October 22, 1983.

[21] National Hemophilia Foundation. AIDS and hemophilia: your questions answered. Hemophilia Information Exchange, AIDS Update/September 1984.

[22] Ratnoff OD, Lederman MM, Krill Jr C, Pass LM, Ross CE. Hemophilia and the acquired immunodeficiency syndrome. Ann Intern Med 1985;102:412.

[23] Gjerset GF, McGrady G, Counts RB, et al. Lymphadenopathy-associated virus antibodies and T cells in hemophiliacs treated with cryoprecipitate or concentrate. Blood 1985;66:718–20.

[24] Desforges JF. AIDS and preventive treatment in hemophilia. N Engl J Med 1983;308:94–5.

[25] Hemophilia Information Exchange. Medical Bulletin #14, Chapter Advisory #19. NHF reaffirms position that product withdrawal should not change use of clotting factor; October 8, 1984.

[26] Smith ES. Priorities for heat-treated antihemophilic factor. Letter to Hemophilia Center Directors and Hematology Consultants; May 24, 1983.

[27] National Hemophilia Foundation. MASAC minutes. Memphis, Tennessee; October 22, 1983.

[28] Resnik S. Blood saga. Berkeley, CA: University of California Press; 1999, pp. 129–30.

[29] Hoyer LW. Letter to the Medical and Scientific Advisory Council; Summer 1984.

[30] Ryan J. Letter of October 26, 1984.

[31] Piszkiewicz D, Bourret L, Lieu M, et al. Heat inactivation of human immunodeficiency virus in lyophilized factor VIII and factor IX concentrates. Thromb Res 1987;47:235–41.

[32] Petricciani JC, McDougal JS, Evatt BL. Case for concluding that heat-treated, licensed antihaemophilic factor is free from HTLV-III. Lancet 1985;ii:890–1.

[33] Levy JA, Mitra GA, Wong MF, Mozen MM. Inactivation by wet and dry heat of AIDS-associated retroviruses during factor VIII purification from plasma. Lancet 1985;i:1456.

[34] Rouzioux C, Chamaret S, Montagnier L, Carnelli V, Rolland G, Mannucci PM. Absence of antibodies to AIDS virus in haemophiliacs treated with heat-treated factor VIII concentrate. Lancet 1985;i:271–2.

[35] Colombo M, Mannucci PM, Carnelli V, Savidge GF, Gazengel C, Schimpf K. Transmission of non-A, non-B hepatitis by heat-treated factor VIII concentrate. Lancet 1985;ii:1–4.

[36] Barrowcliffe E, Edwards SJ, Kemball-Cook G, Thomas DP. Factor VIII degradation products in heated concentrates. Lancet 1986;i:1448–9.

[37] Bird AG, Codd AA, Collins A. Haemophilia and AIDS. Lancet 1985;i:162–3.

[38] Bloom AL. Haemophilia and AIDS. Lancet 1985;i:336.

[39] Scott JP. Minutes of the Medical Advisory Council. Hemophilia Foundation of Illinois; October 22, 1984.

[40] National Hemophilia Foundation Medical and Scientific Advisory Council. Recommendations concerning AIDS and therapy of hemophilia. Medical Bulletin #15, Chapter Advisory #20; revised October 13, 1984.

[41] National Hemophilia Foundation. Revised recommendations concerning AIDS and the treatment of hemophilia. Hemophilia Information Exchange. Medical Bulletin #21, Chapter Advisory #26; April 12, 1985.

[42] National Hemophilia Foundation. On the treatment of infants and young children with severe hemophilia: a follow-up. Hemophilia Information Exchange, Medical Bulletin #25; June 18, 1985.

[43] Anonymous. Blood transfusion, haemophilia, and AIDS. Lancet 1984;ii:1433–5.

[44] Hemophilia Information Exchange, AIDS Update, May 8, 1985. Letter from Rodell MB, Vice President. Regulatory and technical affairs, Armour Pharmaceutical Company; May 1, 1985.

[45] Hemophilia Information Exchange, AIDS Update, June 21, 1985. Notification of market withdrawal by Holst SL, Director of Regulatory Affairs, Hyland Therapeutics Division, Travenol Laboratories, Inc.; June 11, 1985.

[46] Schimpf K, Brackmann HH, Kreuz W, et al. Absence of anti-human immunodeficiency virus types 1 and 2 seroconversion after the treatment of hemophilia A or von Willebrand's disease with pasteurized factor VIII concentrate. N Engl J Med 1989;321:1148–52.

[47] National Hemophilia Foundation. Revised recommendations concerning AIDS and the treatment of hemophilia. Hemophilia Information Exchange, Medical Bulletin #21, Chapter Advisory #26; April 12, 1985.

[48] Lusher J. Letter to colleagues in Region IV; March 13, 1985.

[49] National Hemophilia Foundation. MASAC revises product withdrawal policy in response to new data on heat treatment; product withdrawal announced. Hemophilia Information Exchange, Medical Bulletin #22, Chapter Advisory #27; May 8, 1985.

[50] Laurence J, Brun-Vezinet F, Schutzer SE, et al. Lymphadenopathy-associated viral antibody in AIDS-Immune correlations and definition of a carrier state. N Engl J Med 1984;311:1269–73.

[51] Anonymous. Facts about AIDS: Public Health Service, U.S. Department of Health and Human Services; August 1985.

[52] Anonymous. Update: acquired immunodeficiency syndrome (AIDS) in persons with hemophilia. MMWR Morb Mortal Wkly Rep 1984;33:589–91.

[53] Cameron P. AIDS in haemophiliacs and homosexuals. Lancet 1986;i:36.

[54] Norman C. AIDS trends: projections from limited data. Science 1985;230:1018–21.

[55] Pitchenik AE, Shafron RD, Glasser RM, Spira TJ. The acquired immunodeficiency syndrome in the wife of a hemophiliac. Ann Intern Med 1984;100:62–5.

[56] Kreiss JK, Kitchen LW, Prince HE, Kasper CK, Essex M. Antibody to human T-lymphotropic virus type III in wives of hemophiliacs. Ann Intern Med 1985;102:623–6.

[57] Ragni MV, Kiernan S, Cohen B, et al. Acquired immunodeficiency syndrome in the child of a haemophiliac. Lancet 1985;i:133–5.

[58] Centers of Disease Control (CDC). Recommendations for assisting in the prevention of perinatal transmission of human T-lymphotropic virus type III/lymphadenopathy-associated virus and acquired immunodeficiency syndrome. MMWR Morb Mortal Wkly Rep 1985;34:721–32.

[59] Anonymous. HIV infection and pregnancies in sexual partners of HIV-seropositive hemophilic men-United States. MMWR Morb Mortal Wkly Rep 1987;36:593–5.

[60] Ragni MV, Spero JA, Bontempo FA, Lewis JH. Recurrent infections and lymphadenopathy in the child of a hemophiliac: a survey of children of hemophiliacs positive for human immunodeficiency virus antibody. Ann Intern Med 1986;105:886–7.

[61] National Hemophilia Foundation. The potential spread of HTLV-III to sexual partners of persons with hemophilia. Hemophilia Information Exchange, Medical Bulletin #23; May 10, 1985.

[62] Ragni MV, Gupta P, Rinaldo CR, Kingsley LA, Spero JA, Lewis JH. HIV transmission to female sexual partners of HIV antibody-positive hemophiliacs. Public Health Rep 1988;103:54–8.

[63] CDC Public Health Service guidelines for counseling and antibody testing to prevent HIV infection and AIDS. MMWR Morb Mortal Wkly Rep 1987;36:509–15.

[64] National Hemophilia Foundation. Centers for Disease Control (CDC) reports survey findings on HIV infection and pregnancies in sex partners of HIV seropositive hemophilic men- United States. Hemophilia Information Exchange, Medical Bulletin #56, Chapter Advisory #61; September 14, 1987.

[65] Brettler DB, Forsberg AD, Levine PH, Andrews CH, Baker S, Sullivan JL. HIV isolation studies and antibody testing in household contacts and sexual partners of persons with hemophilia. Arch Intern Med 1988;148:1299–301.

[66] Jackson JB, Sannerud KJ, Hopsicker JS, Kwok SY, Edson JR, Balfour Jr. HH. Hemophiliacs with antibody against human immunodeficiency virus are actively infected. JAMA 1988;260:2236–9.

[67] Lederman MM, Ratnoff OD, Schacter B, Shoger T. Impaired cell-mediated immunity in hemophilia. II. Persistence of subclinical immunodeficiency and enhancement of natural killer activity by lymphokines. J Lab Clin Med 1985;106:197–204.

[68] deShazo RD, Daul CB, Andes WA, Bozelka BE. A longitudinal immunologic evaluation of hemophiliac patients. Blood 1985;66:993–8.

[69] Brettler DB, Forsberg AD, Brewster F, Sullivan JL, Levine PH. Delayed cutaneous hypersensitivity reactions in hemophiliac subjects treated with factor concentrate. Am J Med 1986;81:607–11.

[70] Goldsmith JM, Kalish SB, Green D, Chmiel JS, Wallemark C-B, Phair JP. Sequential clinical and immunologic abnormalities in hemophiliacs. Arch Intern Med 1985;145:431–4.

[71] Goedert JJ, Biggar RJ, Weiss SH, et al. Three-year incidence of AIDS in five cohorts of HTLV-III infected risk group members. Science 1986;231:992–5.

[72] Phillips AN, Lee CA, Elford J, et al. Serial CD4 lymphocyte counts and development of AIDS. Lancet 1991;337:389–92.

[73] Goedert JJ, Kessler CM, Aledort LM, et al. A prospective study of human immunodeficiency virus type I infection and the development of AIDS in subjects with hemophilia. N Engl J Med 1989;321:1141–8.

[74] Goldsmith JM, Deutsche J, Tang M, Green D. CD4 cells in HIV-1 infected hemophiliacs: effect of factor VIII concentrates. Thromb Haemost 1991;66:415–19.

[75] De Biasi R, Rocino A, Miraglia E, Mastrullo L, Quirino AA. The impact of a very high purity factor VIII concentrate on the immune system of human immunodeficiency virus-infected hemophiliacs: a randomized, prospective, two-year comparison with an intermediate purity concentrate. Blood 1991;78:1919–22.

[76] Letter from Donald S Goldman to Robert Morgan, Esq. re: Gallagher V. Cutter; September 14, 1984.

[77] Weinberg PD, Hounshell J, Sherman LA, et al. Legal, financial, and public health consequences of HIV contamination of blood and blood products in the 1980s and 1990s. Ann Intern Med 2002;136:312–19.

[78] National Hemophilia Foundation. Stabilization of the incidence of AIDS in hemophilia reported. Hemophilia Information Exchange, Chapter Advisory #29, Medical Bulletin #24; May 17, 1985.

[79] Koerper MA, Kaminsky LS, Levy JA. Differential prevalence of antibody to AIDS-associated retrovirus in haemophiliacs treated with factor VIII concentrate versus cryoprecipitate: recovery of infectious virus. Lancet 1985;i:275.

[80] Anonymous. Changing patterns of acquired immunodeficiency syndrome in hemophilia patients-United States. MMWR Morb Mortal Wkly Rep 1985;34:241–3.

[81] Goedert JJ, Sarngadharan MG, Eyster ME, et al. Antibodies reactive with human T cell leukemia viruses in the serum of hemophiliacs receiving factor VIII concentrate. Blood 1985;65:492–5.

[82] Ludlam CA, Steel CM, Cheingsong-Popov R, et al. Human T-lymphotropic virus type III (HTLV-III) infection in seronegative haemophiliacs after transfusion of factor VIII. Lancet 1985;ii:233–6.

[83] Rouziouz C, Brun-Vezinet F, Courouce AM, et al. Immunoglobulin G antibodies to lymphadenopathy-associated virus in differently treated French and Belgian hemophiliacs. Ann Intern Med 1985;102:476–9.

[84] AIDS-Hemophilia French Study Group Immunologic and virologic status of multitransfused patients: role of type and origin of blood products. Blood 1985;66:896–901.

[85] Ragni MV, Winkelstein A, Kingsley L, Spero JA, Lewis JH. 1986 update on HIV seroprevalence, seroconversion, AIDS incidence, and immunologic correlates of HIV infection in patients with hemophilia A and B. Blood 1987;70:786–90.

[86] Anonymous. Update: acquired immunodeficiency syndrome-Europe. MMWR Morb Mortal Wkly Rep 1985;34:21–31.

[87] Gurtler LG, Wernicke D, Eberle J, Zoulek G, Deinhardt F, Schramm W. Increase in prevalence of anti-HTLV III in hemophiliacs. Lancet 1984;ii:397–8.

[88] Anonymous. Safer factor VIII and IX. Lancet 1986;ii:255–6.

[89] Chorba TL, Holman RC, Strine TW, Clarke MJ, Evatt BL. Changes in longevity and causes of death among persons with hemophilia A. Am J Hematol 1994;45:112–21.

[90] White II GC, McMillan CW, Kingdon HS, Shoemaker CB. Use of recombinant antihemophilic factor in the treatment of two patients with classic hemophilia. N Engl J Med 1989;320:166–70.

[91] Schwartz RS, Abildgaard CF, Aledort LM, et al. Human recombinant DNA-derived antihemophilic factor (factor VIII) in the treatment of hemophilia A. N Engl J Med 1990;323:1800–5.

[92] Hoxie JA, Rackowski JL, Cedarbaum AJ, Hurwitz S, Catalano PM. Relation of human T lymphotrophic virus type III antibodies to T lymphocyte subset abnormalities in hemophiliac patients. Am J Med 1986;81:201–7.

[93] 543 of 2764 (19.6%); ATHN Research Report Brief, December 31, 2013. As of March 31, 2015, 663 of 3603 (18.4%) persons with hemophilia are HIV-positive.

[94] Gibbons J, Cory JM, Hewlett IK, Epstein JS, Eyster ME. Silent infections with human immunodeficiency virus type 1 are highly unlikely in multitransfused seronegative hemophiliacs. Blood 1990;76:1924–6.

[95] Lane J, McLaren PJ, Dorrell L, et al. A genome-wide association study of resistance to HIV infection in highly exposed uninfected individuals with hemophilia A. Hum Mol Genet 2013;22:1903–10.

[96] Merigan TC, Amato DA, Balsley J, et al. Placebo-controlled trial to evaluate Zidovudine in treatment of human immunodeficiency virus infection in asymptomatic patients with hemophilia. Blood 1991;78:900–6.

[97] Monroe KM, Yang Z, Johnson JR, et al. IFI16 DNA sensor is required for death of lymphoid CD4 T cells abortively infected with HIV. Science 2014;343:428–32.

Chapter 8

The Institute of Medicine Study

In 1995, the US Institute of Medicine (IOM) appointed a committee to study human immunodeficiency virus (HIV) transmission through blood and blood products. They published a report entitled *HIV and the Blood Supply: An Analysis of Crisis Decisionmaking* [1]. The report was several hundred pages in length and included a history of the epidemic, the analysis and conclusions of the Committee, and a set of recommendations for the US Public Health Service, the Centers for Disease Control (CDC), and the Food and Drug Administration (FDA). There were several suggestions for improving communications between Government agencies, physicians, the National Hemophilia Foundation (NHF), and patients.

The chapter on donor screening and deferral (chapter 5 of the Report) is particularly relevant to the safety of blood products. In January 1983, the CDC suspected that a transmittable agent was present in the blood supply and suggested that donors should be questioned about risk behaviors and blood be tested for hepatitis antibody. The blood services community resisted these measures, and they were not implemented by most blood banks. However, the manufacturers of clotting factor concentrates were more amenable to actions that would protect the users of their products and would avoid future product liability lawsuits. When the FDA urged that plasma from donors at increased risk for acquired immunodeficiency syndrome (AIDS) not be made into clotting factor concentrates, most companies heeded their recommendations. In fact, as early as 1982, one manufacturer excluded donors who had been to Haiti, used IV drugs, or had sexual contact with another man [2], and other companies soon followed their lead.

Blood banks did not implement direct questioning of donors about sexual preferences; instead, they provided written educational materials about AIDS and the necessity for high-risk groups to refrain from donation. They thought that direct questioning would damage donor motivation and be counterproductive to risk reduction. They were also concerned about the propriety of asking donors about sexual activities, the truthfulness of donors responding to such questions, and the legal and political implications of direct questioning. Therefore, blood bank directors did not institute donor screening and blood testing; they stated that the costs of these procedures would outweigh their potential

Linked by Blood: Hemophilia and AIDS. DOI: http://dx.doi.org/10.1016/B978-0-12-805302-7.00008-2

benefits. Although the FDA had the authority to compel blood banks to adopt these procedures, they did not issue specific recommendations until almost a year after the CDC called for donor questioning and blood testing.

The IOM Committee concluded that the lack of consensus about the costs and benefits of the various donor screening procedures limited the responses among organizations to the issues of donor safety. Blood bank directors did not want to assume the expenses of new screening measures, and they did not wish to antagonize gay organizations. Government agencies were reluctant to issue new regulations about donor screening for a number of reasons. First, the FDA was not convinced of the credibility of the CDC's warnings about contamination of the blood supply. Second, the experts who advised the FDA, the Blood Products Advisory Committee, lacked personnel with the training and experience to decide whether the risks of HIV transmission outweighed the risks of further stigmatizing homosexuals. Third, the Reagan administration did not see AIDS as an urgent and serious public health threat, and it was generally opposed to issuing new regulations. In the absence of any serious commitment to change the status quo, nothing was done. The IOM report states: "When confronted with a range of options for using donor screening and deferral to reduce the probability of spreading HIV through the blood supply, blood bank officials and federal authorities consistently chose the least aggressive option that was justifiable."

Chapter 7 of the Report analyzed risk communication to physicians and patients. It began with a discussion of the options for risk reduction. The Committee thought that heat treatment of clotting factor concentrates should have been developed prior to 1980, as part of an effort to decrease the transmission of hepatitis viruses. Had this been pursued, transmission of HIV, which is more sensitive to heat inactivation than hepatitis virus, might have been decreased. However, the prevailing view at the time was that hepatitis was an "acceptable risk" because of the tremendous benefit of clotting concentrates in controlling bleeding. Furthermore, attempts to inactivate the hepatitis virus were not initiated because of concerns that heat might potentially damage plasma clotting factors.

The range of treatment options available to the Hemophilia Community was limited. The chief among these was the use of cryoprecipitate rather than concentrate. The Committee faulted the National Hemophilia Foundation (NHF) and its Medical and Scientific Advisory Council (MASAC) for dismissing cryoprecipitate as not being feasible without a thorough analysis, and for not discouraging the prophylactic use of the potentially infectious concentrates. In addition, there were no alerts about the possibility that infected hemophiliacs could sexually transmit the disease to their spouses. Perhaps NHF's greatest failing was not recommending the use of heat-treated concentrates when they first became available in mid- 1983; the official MASAC recommendation to switch to a heat-treated concentrate did not come until October 1984, and widespread use was not adopted until 1985.

The Committee examined several case histories showing the failure of communication between the Hemophilia Leadership, patient physicians, and hemophiliacs. In the first case study, a physician with factor IX deficiency hemophilia read a 1982 description of AIDS in the *New England Journal of Medicine*. He recognized that AIDS might be transmitted by blood products and sought advice from hemophilia experts at his treatment center. He was told that the risks were minimal and to adopt a "wait and see" attitude. However, he opted to receive plasma donated by his friends and colleagues (Directed Donation). The local chapter of the Red Cross declined to assist him, but a private blood bank agreed to accept his directed donations and provide him with the fresh-frozen plasma he needed to treat his hemophilia. He never became infected with the HIV virus.

The second case involved a patient with severe hemophilia who had been treated with concentrate since 1974. He became aware of the risk of AIDS from blood products in 1982; in early 1983, he learned of several cases of AIDS in hemophiliacs. Although he was informed by his physicians that the NHF recommended no change in treatment, he began to limit his use of concentrate. Nevertheless, he was found to have immunodeficiency later that year, but was told by his physician that he had only a mild infection. Another physician, when asked for a second opinion, dismissed the diagnosis of AIDS. Despite these assurances, this patient decided to switch from concentrate to cryoprecipitate. After much difficulty, he was finally able to convince the local Red Cross to provide the cryoprecipitate. He continued using this material until heat-treated concentrate became available in 1986; later that year, he began therapy with azidothymidine for his now-confirmed HIV infection. He told the Committee that NHF neglected its responsibilities by inadequately warning patients of the risks of concentrate therapy, and that physicians were misled into thinking the risks were low.

In the third study, a physician involved in hemophilia care described the response of his Treatment Center to the AIDS epidemic. The Center was aware that the disease was transmitted by blood products and that cryoprecipitate was safer than concentrate. However, there was insufficient cryoprecipitate to meet the needs of all patients, and it was more difficult to obtain. Despite these problems, patients were given the option of cryoprecipitate but few decided to use it. Even the option of heat-treated concentrate was not readily accepted for several reasons: the product was more expensive, there were concerns that the heated clotting factor might engender treatment resistance, and it was believed that most patients had already been exposed to the virus. Another factor was the recent bleeding death of a teenager who had a head injury but was afraid to treat himself with concentrate. This led many physicians to question the wisdom of limiting the use of clotting factor. The physician complained that lack of guidance from the FDA and NIH resulted in the absence of a coherent treatment strategy.

Another case study presented a remarkable account of treatment inconsistency. The father of two boys with hemophilia reported that his older son was treated with concentrate while the younger, born in early 1982, received cryo-precipitate because the hematologist said that there were too many unknowns about concentrate. When the father asked why the older boy was still being treated with concentrate, he was told that the NHF was recommending that there should be no changes in treatment already initiated. In August 1983, the father received a booklet from a concentrate manufacturer reporting that their product was a potential source of AIDS. At that point, he again wondered why both of his sons were not treated with cryoprecipitate. The older boy subsequently was found to be HIV-positive, but the younger son never became infected.

The last case presented was that of a pregnant woman who received 7 pints of blood in 1980 when she hemorrhaged during childbirth. She breastfed her infant and, in 1984, she gave birth to a second child. The first child became ill in September 1985, but there was a considerable delay until it was recognized that this child had AIDS; at that point, the woman and her second child were tested and both were discovered to be HIV-positive. It was suspected that HIV was present in one or more of the pints of blood transfused in 1980, and that the virus was transmitted to the woman's firstborn by her breast milk and was transmitted *in utero* to the second child. When it became known that the family had HIV, friends stopped visiting and the children were denied attendance at summer camp. Despite this social isolation, the woman became active in raising funds for AIDS research and helped to establish the Pediatric AIDS Foundation. She and her first child eventually died of AIDS.

The IOM Committee reported these cases to demonstrate the limited clinical options available to physicians and patients confronted with the increasing risks of blood and blood products. The case scenarios showed that authorities often underestimated the dangers of transfusion and gave unjustified assurances about the safety of therapeutic materials. Even after the link between concentrates and AIDS was recognized, blood banks and the Red Cross were reluctant to support directed donations or provide cryoprecipitate. Some physicians assumed that most of their hemophilia patients were already infected and that switching to cryoprecipitate would provide suboptimal treatment for bleeding; they used this logic to justify the continued prescription of clotting factor concentrates. The Committee noted that this issue was infrequently discussed with patients, who might have viewed AIDS as a much worse outcome than an inadequately treated hemorrhage. Full disclosure of the risks and benefits of continued concentrate use would have allowed patients to exercise the right of self-determination. This is consistent with a central tenet of medical practice known as the doctrine of informed consent, which requires physicians to disclose to patients medical information relevant to making treatment decisions. In 1981, the Judicial Council of the American Medical Association wrote: "The patient's right of self-decision can be effectively exercised only if the patient possesses enough information to enable an intelligent choice… Social policy does not accept the paternalistic view that the physicians may remain silent because divulgence might prompt the patient to forego medical therapy." Doctors are obligated to provide patients with whatever information is available; they can suggest treatment options, but the final decision rests with the patient.

The NHF (and its MASAC) was the only agency with the expertise to act as an intermediary between federal agencies and the hemophilia community. It became a clearinghouse for information about HIV and AIDS. Although the NHF might make recommendations to local chapters about management issues, there was no mechanism to ensure that the information was passed to chapter members. The absence of a National Patient Registry meant that there was no way to communicate with all the country's hemophiliacs. Furthermore, many patients were under the care of physicians not associated with the Hemophilia Treatment Centers. Finally, hemophiliacs using home treatment visited their physicians and centers infrequently. In addition to these logistical difficulties, the failure of NHF and MASAC to alert hemophiliacs to the dangers of AIDS was attributed to:

- Tendencies to downplay the risks of AIDS
- Overconfidence in the benefits of concentrates
- Belief that hepatitis was a medically acceptable risk
- Reluctance to impart "bad news" to patients
- Problems communicating uncertainty

A discussion of treatment alternatives and risks is a central tenet of medical practice, and the NHF and its physician advisory board should have made such information available to the members of the hemophilia community. That would have enabled hemophiliacs and their physicians to make informed decisions about whether to continue concentrates or switch to cryoprecipitate (factor VIII–deficient patients) or fresh-frozen plasma (factor IX–deficient patients). The IOM Committee's Report noted that in 1983, the NHF made a policy decision to retain tight control of the information distributed to its members, to allay public fears about AIDS, and to emphasize that clotting factor use should be maintained. This continued advocacy of concentrate use by the NHF did not empower patients to make their own treatment choices and was an institutional failure of communication.

The Committee's report suggested actions that might have limited the dissemination of the virus. They may be summarized as follows:

1. Earlier institution of automatic withdrawal of concentrate lots containing blood from donors suspected of having AIDS, and switch from concentrate to cryoprecipitate in mild to moderate hemophilia.
2. Heat treatment of concentrate should have been developed prior to 1980; the technology was available but there were few incentives for plasma fractionation companies to implement production.
3. Institution of donor screening and exclusion of those with a positive test result for hepatitis or men having sex with men would have improved the safety of the blood supply.

4. The information provided by NHF would have benefitted from input by a wider range of scientific and medical experts, particularly those with training in infectious diseases.
5. Failure of institutional leadership to communicate the growing concerns about the contamination of blood products.

The report concluded with 14 recommendations. With regard to blood safety, the IOM Committee called for the appointment of a Blood Safety Director and Council charged with addressing current and future threats to the blood supply. Council members would propose strategies to overcome these threats and monitor their implementation. In addition, they would alert scientists to the needs and opportunities for research to maximize the safety of blood and blood products. The Council would assume responsibility for educating public health officials, clinicians, and the public at large about the nature of threats to the blood supply and the strategies for dealing with such threats.

The IOM Committee also recommended establishing a no-fault compensation system for persons suffering injury from blood or blood products, possibly financed by a tax or fee paid by manufacturers or recipients of these therapies. Another recommendation was that the CDC should have a surveillance system to detect, monitor, and warn of adverse effects due to blood and blood products. There were also several recommendations that would improve the ability of the FDA to regulate the blood products industry. Finally, it was suggested that voluntary organizations should avoid conflicts of interest by selecting impartial experts for panels providing medical advice.

ANALYSIS

The Report of the IOM Committee to Study HIV Transmission Through Blood and Blood Products was highly critical of blood procurement and processing procedures and suggested several actions that might have limited the dissemination of the virus. In retrospect, each of these actions would have been highly effective. But were they feasible in the early 1980s? Certainly, early in the epidemic the FDA could have issued a strong recommendation that manufacturers should avoid collecting blood from donors at high risk for AIDS. Companies had already discontinued collecting blood from prisoners, and excluding another high-risk population would not have been excessively burdensome. With regard to product quality improvement, the Industry was aware in the 1970s that their concentrates transmitted infection with hepatitis viruses. However, they also knew that clotting factor VIII is present in very low concentrations in plasma and its activity is lost during purification procedures. Perhaps manufacturers had investigated the feasibility of preparing heated concentrates but did not institute commercial production until they were convinced that there was a market for such a product. Pressure by the NHF and governmental agencies might have accelerated development and earlier availability of these safer concentrates.

Although converting from concentrates to cryoprecipitate would have decreased the spread of HIV, this suggestion fails to recognize that cryoprecipitate is suitable for only a minority of hemophiliacs. It does not benefit the 20% with factor IX deficiency, and its use is impractical for the majority of persons with severe factor VIII deficiency. This is because cryoprecipitate is difficult to infuse and store in the home setting, which is where most hemophiliacs receive their treatment. Furthermore, the amount of cryoprecipitate required to control and prevent bleeding exceeds what most blood banks can provide. Another disadvantage of cryoprecipitate is that it contains a variety of proteins in addition to factor VIII. Allergies to these proteins may cause hives, wheezing, and other unpleasant reactions. These problems with cryoprecipitate explain the preference of hemophiliacs for the more highly purified concentrates. Therefore, only those with mild or moderately severe factor VIII hemophilia who infrequently used blood products would have benefitted from a switch from concentrate to cryoprecipitate.

Could the NHF and its medical advisory board have communicated more clearly the risks of AIDS to its membership? The Commission faulted the leadership for its persistence in recommending that members should continue to use concentrate despite knowing that AIDS had been diagnosed in some donors and products were being recalled. The IOM also noted that NHF and its physician advisors had financial conflicts of interest that might have influenced their recommendations. In addition, although the doctors advising NHF were experts in hemophilia care, they were not especially knowledgeable about infectious diseases. Transfusion-transmitted infections such as hepatitis were already widespread in hemophiliacs, and it would have been prudent to appoint infectious disease specialists to physician advisory boards.

However, the NHF probably assumed that their members had doctor–patient relationships with their personal physicians, as is customary in US health care. They might have believed that these physicians were current with best medical practices, a competency attained by reading the medical literature, attending didactic courses, and consulting with experts. It was the duty of these physicians, as well as the NHF, to provide advice about treatment. Therefore, personal physicians and regional hemophilia center staff bear some responsibility for failures of communication.

In retrospect, a perplexing question is why physicians and patients continued to use concentrates in 1983 and 1984 despite the accumulating evidence that they transmitted AIDS. Hughes-Jones suggests this conundrum can be best understood in terms of risk analysis, that is, by examining how benefits and harms were perceived at the time [3]. For example, in the 1960s it was recognized that transfusions carried a risk of hepatitis of between 1% and 10%, but this harm was considered tolerable. In the 1980s, practitioners were well-aware of the benefits of concentrates but completely underestimated the danger of HIV infection.

One of the greatest failings of NHF and its physicians advisors was the lack of appreciation of the devastating effect that AIDS was having on the homosexual community. In the 1980s, sex was generally a taboo subject. Physicians infrequently discussed sexual issues and avoided mention of homosexuality in particular. As noted by Shilts, happenings in the gay community were rarely reported by the media [4]. If NHF and its physicians had been aware of the rapid spread of the AIDS virus among the gay population, its long latency period, and its often fatal outcome, then they might not have been so certain that HIV would infrequently affect hemophiliacs and not increase their risk of getting deadly infections; and they might have been more open to prescribing alternative treatments. The taboos of explicit discussions of sex might also explain why NHF failed to warn its members about the risk of sexually transmitting HIV infection to their spouses.

In summary, the IOM Committee faulted physicians and organizations for their failure to alert hemophiliacs to the risks of HIV transmission by blood and blood products. Many doctors did not adequately apprise their patients about the scientific uncertainties regarding the safety of blood products, and they avoided discussing the dire prognosis of AIDS. Although the NHF provided access to information about AIDS as soon as it became available, it failed to fully present the risks and benefits of its recommendations, especially in the latter years of the epidemic when it became clear that clotting factor concentrates transmitted HIV. At that stage of the epidemic, NHF should have warned its membership about the potential risk that infected individuals posed to others through sexual contact. Finally, there was a lack of effective leadership at Government agencies that resulted in less than effective donor screening, weak regulatory actions, and inadequate communication about the threat of AIDS.

KEY POINTS: THE IOM COMMITTEE REPORT

1. In January 1983, the CDC strongly suspected that AIDS was transmitted by blood, but the FDA failed to make explicit recommendations about donor screening.
2. The FDA failed to encourage manufacturers to develop heat-treated clotting factor concentrates.
3. A primitive communication network and absence of a patient registry were serious obstacles that prevented the hemophilia leadership from reaching many hemophiliacs in the community.
4. The continued advocacy of concentrate use might have been influenced by conflicts of interest affecting NHF and its medical advisory board. Furthermore, the paternalistic attitude of NHF and MASAC did not encourage members to make their own treatment decisions.
5. The IOM Committee recommended that a Blood Safety Director and Blood Safety Council be established whose role would be to assess threats to the blood supply; communicate this information to public health officials, clinicians, and the public; propose strategies to overcome such threats; monitor their implementation; and alert scientists of research opportunities to improve blood safety.

6. Other recommendations were that the CDC should institute a surveillance system to detect, monitor, and warn of adverse effects of blood and blood products, and the Government should establish a no-fault compensation system for individuals injured by these products.
7. Lastly, Voluntary organizations should avoid conflicts of interest by selecting impartial experts for panels providing medical advice.

REFERENCES

[1]   Leveton LB, Sox Jr. HC, Stoto MA, editors. HIV and the blood supply: an analysis of crisis decisionmaking. Washington, DC: National Academy Press; 1995.
[2]   Gury DJ. Alpha Therapeutic Corporation letter sent by Federal Express to all plasma source affiliates; December 17, 1982.
[3]   Hughes-Jones NC. Risk assessment and factor VIII concentrates. Lancet 1995;345:502–3.
[4]   Shilts R. And the band played on. New York, NY: St Martin's Press; 1987.

Chapter 9

A Summary of Factors That Enabled the Epidemic

The central question addressed by this book is why hemophiliacs were so vulnerable to human immunodeficiency virus (HIV) infection, and the simplest answer is that hemophilia and acquired immune deficiency syndrome (AIDS) are linked by blood. HIV-infected blood contaminated the plasma pools used for manufacturing the clotting factor concentrates that were prescribed to control hemophilic bleeding. The sources of the virus were HIV-infected blood donors, the blood banks that accepted their blood, and the pharmaceutical companies that used the tainted plasma to prepare the concentrates. Widespread dissemination of the virus occurred because physicians steadfastly encouraged the use of the concentrates and the National Hemophilia Foundation (NHF) urged its members to follow this medical advice, even though there was considerable evidence that the concentrates were contaminated. Finally, several government agencies—the Centers for Disease Control (CDC), the Office of Biologics, and the Food and Drug Administration (FDA)—failed to issue directives that would have limited the spread of the virus.

Our blood supply is dependent on volunteer donors who give their blood so that others might regain their health. The HIV-infected persons who donated blood were not aware that they were ill or had a communicable disease because they were asymptomatic during the long incubation period from HIV infection to clinical illness (AIDS). So, one cannot fault the blood donors who altruistically gave their blood to help those in need of this invaluable commodity.

Blood bank directors maintained throughout the first years of the epidemic that blood was safe because only one in a million donors had AIDS. This low frequency was used to justify their reluctance to screen donors for AIDS risk factors and to test the blood for surrogate markers of HIV infection. They supported this position by noting that screening procedures might discourage potential donors, and that blood testing would increase the cost of blood processing. These arguments convinced the FDA that mandating new procedures to safeguard the nation's blood supply was unnecessary.

The main premise of the blood bank directors was that HIV contamination of the blood supply was a rare event, and this was based on the fact that signs of

Linked by Blood: Hemophilia and AIDS. DOI: http://dx.doi.org/10.1016/B978-0-12-805302-7.00009-4

AIDS were very infrequent in those who had recently donated blood. However, the number of persons with symptoms could be just the tip of the iceberg of the many harboring the infection but remaining asymptomatic. Blood bankers should have been aware of this possibility because they knew that the hepatitis B virus acted in this fashion, and they had begun to test donors for this virus as early as 1971. When the CDC warned in 1982 that there might be a lag period of a year or more between the time of exposure to the causative agent of AIDS and the onset of symptoms, blood bank directors should have suspected that the number of donors with AIDS might underestimate the number with asymptomatic HIV infection. In addition, it was known that many of the donors were homosexuals and that AIDS was rife in the homosexual community. This was another reason to institute donor screening and testing for AIDS risk factors. Failure to initiate such testing between 1983 and 1985 was the main contributor to the epidemic; testing should have been made mandatory by government regulators.

Additionally, plasma collected by blood banks and other collection facilities was pooled and used to prepare clotting factor concentrate. The pooling process posed the risk that infected material from one donation could contaminate the entire production. Recognition of this possibility should have prompted greater efforts to sterilize the final product. Prior to 1980, manufacturers had considered additional plasma processing to eliminate hepatitis viruses but did not pursue this avenue. As the Institute of Medicine Committee Report stated, manufacturers should have instituted procedures to sterilize plasma, even though there were theoretical reasons to believe that such measures might partially inactivate factor VIII. At the very least, the sterile concentrate could have been used to treat patients with factor IX–type hemophilia. As it turned out, it was possible to produce a heat-sterilized concentrate that retained factor VIII activity, and this became generally available early in 1985. The heated concentrate does not transmit HIV, but there is still concern that other microbes that are resistant to heating might one day contaminate plasma pools. This fear about the future safety of plasma products has led to a preference for clotting factors made by recombinant DNA technology, a process that does not require human plasma. As a consequence of the HIV epidemic, it is now accepted that all products used for the treatment of disease should be of the highest quality, and it is the responsibility of government regulators to ensure the compliance of manufacturers with these standards.

The NHF was organized by persons with hemophilia and their family members to provide better treatment for bleeding disorders. This goal was to be accomplished by gathering information from experts and communicating it to the membership. The quality of the information provided was dependent on the expertise of the physician advisors as well as their ability to impart their knowledge to the NHF Board. The members of NHF's Medical and Scientific Advisory Council (MASAC) were directors of federally funded Hemophilia Treatment Centers, and the basic tenet of the comprehensive care they provided was the early use of clotting factor concentrates to prevent and treat hemorrhages.

This approach had been effective in preventing disability, enhancing quality of life, and extending longevity in persons with hemophilia. Therefore, it is not surprising that these physicians steadfastly rejected any suggestions that the use of concentrate, even if potentially HIV-infected, should be limited or abandoned. Perhaps they were unaware that HIV infection, in contrast to hepatitis, was frequently fatal. In 1982, the CDC reported a mortality rate of 40% in homosexuals with AIDS; this information was published in the *Morbidity and Mortality Weekly Report* and was available to all physicians who read this periodical. So, when the risks and benefits of clotting factor concentrates were categorized (see chapter 5, page 62), the 40% chance of death from HIV infection should have been included in the column showing the risks of using blood products. This might have influenced the decisions of patients and physicians to continue using potentially contaminated concentrates.

Hemophilia caregivers did have the option of consulting with infectious disease specialists who were more knowledgeable about AIDS, and these experts might have recommended restricting exposure to blood products. However, such physicians were not represented on the NHF Board. Without their input, the information received and disseminated by the NHF was deficient, as noted in chapter 5. There were no warnings given about the growing contamination of blood products and no evidence to support their assurances about the safety of concentrates. Furthermore, the NHF's concerns that heated concentrates might be deleterious were unsubstantiated and led to a delay in the implementation of these safer blood products. Downplaying the risks of HIV infection and underestimating its dire consequences led the NHF "experts" to continue advising hemophiliacs not to alter their use of concentrates.

The recommendations of NHF and its medical advisory committee (MASAC) might also have been affected by monetary considerations. Concentrate manufacturers contributed to NHF programs and provided grant support and honoraria to some members of MASAC. In addition, the consortia described in chapter 2, page 20, purchased and distributed commercial clotting factor concentrates, and the income from this activity was used to support salaries and administrative expenses. It was in the financial interest of these physicians and organizations for concentrate use to be maintained rather than adopting the alternative of recommending cryoprecipitate or avoiding blood products altogether.

The close relationships between the NHF and industry were often subtle. Susan Resnik recalls that when she worked for the NHF, she promoted an educational guide that contained a self-infusion home therapy module [1]. She subsequently realized that this guide encouraged the use of pharmaceutical products and was really a marketing tool for the concentrate manufacturers. This was just one of the ways that commercial interests could promote their products under the guise of "educating consumers." Another strategy used by industry was the employment of persons with hemophilia as "sales representatives." This ensured that these individuals would use company products and recommend them to other persons with hemophilia. Thus, several of the consumer and physician representatives on the board of NHF were "enmeshed with indus-try" [2]. The continued exhortation to "maintain the use of concentrate" [3] late in the course of the AIDS epidemic might have been biased by these industry relationships and represented a conflict of interest.

Government agencies were also remiss. The Institute of Medicine report identified poor leadership and excessive bureaucracy (see chapter 8, page 96). Federal regulators failed to demand that blood banks institute screening pro-cedures and test the donated blood; they also did not do enough to incentivize manufacturers to make a safer product. In addition, there was a lack of coop-eration among the several agencies that might have contributed to the failure to take action. In the years following the epidemic, the negligence of govern-ment bureaucrats led to considerable litigation; although officials in the United States escaped prosecution, in other countries criminal charges were brought against administrators for not communicating risks to hemophiliacs (Japan) and for delaying the removal of contaminated concentrates from pharmacies (France) [4].

One of the major failings in the early 1980s was that the larger society ignored AIDS as it was attacking members of the gay community [5]. There was little coverage by the media and research was underfunded by government. Disease epidemics often affect one segment of the population more than others. For example, children are more susceptible to polio, and the elderly are more susceptible to Legionnaire's disease. The AIDS epidemic was no exception; it centered mainly on homosexuals and hemophiliacs. The response of the general public was that AIDS was "their" problem, and persons with the disease were treated with hostility. Hemophiliacs were shunned by neighbors and employers, and school boards suspended infected boys. There was no medical justifica-tion for these actions; in fact, there were no instances of transmission of HIV to siblings of infected boys and no spread of the virus in nursery or boarding schools attended by HIV-positive hemophiliacs (see chapter 6, page 74). There were also financial repercussions. HIV-positive men with hemophilia lost their jobs but still had to find the money for frequent clinic visits and occasional hospitalizations. In addition, the more highly purified concentrates were more expensive. This financial burden was partly relieved by payments from concen-trate manufacturers, and some treatment expenses were underwritten by State Programs; however, over the ensuing years, most hemophiliacs have accumu-lated considerable debt.

KEY POINTS

- Hemophiliacs became infected because HIV-infected blood contaminated the plasma pools used for manufacturing clotting factor concentrates.
- Delays in implementing screening procedures for blood donors were justi-fied by the assumption that AIDS was a rare event that did not warrant the projected costs.

- Pharmaceutical companies were reluctant to adopt new procedures for sterilizing concentrates because of fears that the clotting factors would be degraded or diminished.
- The advice that the consumer organization (the NHF) received from its professional members was inadequate and possibly influenced by conflicts of interest.
- Government regulators failed to demand that blood bank directors and concentrate manufacturers improve the safety of their products.
- Prejudice against gays and irrational fears of contagion led to the social and economic isolation of infected individuals and their families.

REFERENCES

[1]   Resnik S. Blood saga. Berkeley, CA: University of California Press; 1999, p. 196.
[2]   Resnik S. Blood saga. Berkeley, CA: University of California Press; 1999, p. 129–30
[3]   Hemophilia Information Exchange. AIDS update, Medical Bulletin #20; December 13, 1984.
[4]   Weinberg PD, Hounshell J, Sherman LA, et al. Legal, financial, and public health consequences of HIV contamination of blood and blood products in the 1980s and 1990s. Ann Intern Med 2002;136:312–19.
[5]   Shilts R. As the band played on. New York, NY: St Martin's Press; 1987, p. 630.

Chapter 10

Contributions That Mitigated the Epidemic

The transmission of human immunodeficiency virus (HIV) to hemophiliacs was finally brought under control by three major medical advances: the development of a simple test for HIV; the sterilization of clotting factor concentrates; and the discovery of antiretroviral medications.

A sensitive and specific test for HIV (originally called HTLV-III) was described by investigators from the National Cancer Institute in 1983 [1]. This assay revealed the presence of antibodies to the virus in 100% of patients with acquired immunodeficiency syndrome (AIDS) [2]. The prevalence of HIV antibodies in hemophiliacs was found to be 74% for those with factor VIII deficiency and 39% for those with factor IX deficiency [3]. The availability of a test for HIV was essential for the diagnosis of infected blood donors, hemophiliacs, and others, as well as for the identification of contaminated blood products.

Although other companies had experimented with methods to sterilize clotting factor concentrates, Hyland Laboratories was the first to announce the development of a heat-treated product. This concentrate was licensed by the Food and Drug Administration (FDA) in 1983; it provided partial protection against the transmission of hepatitis viruses. A second heat-treated product was introduced by Cutter Laboratories in 1984; it was manufactured by a procedure that was extremely effective in eliminating HIV [4]. The epidemic of new HIV infections in hemophiliacs ended when these concentrates came into general use in the summer of 1985.

The first safe and effective antiretroviral drug was azidothymidine (AZT), developed by the British company Burroughs Wellcome. Treatment with AZT for 8 weeks decreased the blood levels of HIV by 83% [5], and the infections in fewer patients progressed to AIDS [6]. On the basis of these clinical trials, AZT was licensed by the FDA in 1986. This drug was subsequently shown to prolong the lives of hemophiliacs and was a seminal advance.

The years prior to the implementation of these medical discoveries were a time of great anxiety, uncertainty, and grief for the members of the hemophilia community. The tragedy was that people with hemophilia and their physicians blindly followed the recommendations of the National Hemophilia Foundation

Linked by Blood: Hemophilia and AIDS. DOI: http://dx.doi.org/10.1016/B978-0-12-805302-7.00010-0

(NHF) not to change or stop treatment with clotting factor concentrates, despite mounting evidence that these products were the source of AIDS. However, there were a few prescient individuals who recognized the impending disaster and advocated for measures to protect people with hemophilia.

Nathan J. Smith (1941–86) was born with severe hemophilia and suffered from recurrent joint hemorrhages that eventually required replacement of both knees. Despite this handicap, he obtained a college education and found employment with a major insurance company. Mr Smith worked tirelessly to improve the lives of people with hemophilia. His job brought him into contact with hemophilia physicians, and his conversations with these doctors convinced him that HIV infection was going to be a major threat to the hemophilia population. He vigorously campaigned for more research into the causes and treatment of AIDS during meetings with many of the physicians responsible for hemophilia treatment. In the autumn of 1982, he began publishing *Hemophilia Digest*. Its goal was to bring the latest medical and scientific information about bleeding disorders to the hemophilia community. He served as President of the Utah Chapter of NHF, and subsequently as President of the national organization. In 1985, he testified before a US Senate Committee to encourage the early release of Zidovudine (AZT). Nathan Smith was diagnosed with AIDS later that year and died soon thereafter of massive bleeding from a lymphoma.

Bruce Evatt, MD, a staff hematologist at the Centers for Disease Control and Prevention (CDC), was the first physician to warn of the danger that AIDS posed to the hemophilia community. He had received a report in early 1982 describing the death of a hemophiliac from pneumocystis pneumonia [7]. Dr Evatt suspected that the man might have had AIDS based on the accumulating accounts of similar infections in homosexuals. He communicated his concerns to NHF later that year and raised the possibility that a virus might be contaminating commercial clotting factor concentrates. Then, in July 1983, Dr Evatt reported that transfusion of blood and blood products was a source of AIDS. He repeatedly urged blood banks and clotting factor manufacturers to exclude high-risk donors and advocated for the use of heated concentrates. Had his advice been heeded earlier, many individuals would have avoided being infected by HIV.

Another physician who recognized the hazards of commercial clotting factor concentrates was Oscar D. Ratnoff, MD (1916–2008), Professor of Medicine at Case Western Reserve University in Cleveland, OH. He was a brilliant clinician and scientist who made major contributions to knowledge about bleeding and clotting disorders. As noted in chapter 2, page 23, Dr Ratnoff expressed concern that manufacturers were using blood from potentially infected donors to prepare commercial concentrates. He thought that microbes present in these products might infect his patients; therefore, he prescribed only cryoprecipitate for the control of bleeding. Subsequently, he reported that AIDS was more frequent in users of concentrate than cryoprecipitate. He made seminal observations about HIV infection and immunodeficiency in hemophiliacs, and his work was published in influential medical journals. His very vocal concerns about blood product safety convinced manufacturers to be more careful in their selection of plasma donors and spurred efforts to prepare safer clotting factor concentrates.

Carol K. Kasper, MD, a physician at the Los Angeles Orthopedic Hospital, also recognized the need for better communication of the risks of AIDS. In addition to caring for a large number of persons with bleeding disorders, she published *The Hemophilia Bulletin*, a private newsletter for those who treated hemophilia. The January 1983 issue was exclusively devoted to AIDS and hemophilia; Dr Kasper wrote that "Immediate action can be taken to reduce the presumed exposure to the mystery virus in some patients by using cryoprecipitate instead of Factor VIII concentrate...." This sage advice early in the epidemic may have saved lives. In addition, she explored other topics such as alternatives to blood products for the management of hemorrhages and measures to protect nurses and laboratory workers from infection. The *Bulletin* appeared regularly during the years of the epidemic and included summaries of discussions at national and international medical conferences. It was a splendid source of factual material and practical advice for those responsible for hemophilia care.

Hemophiliacs often asserted that because they were major consumers of blood and blood products, they were the bellwethers for the safety of the nation's blood supply. If blood became contaminated, then they would be the first to be infected. When AIDS appeared in their community, hemophiliacs and their families became vocal proponents for increasing public education about the disease and committing more funds for research. They demonstrated great fortitude despite being barred from schools, dismissed from places of employment, and subject to the isolation caused by hostile neighbors. Men and boys infected by HIV bravely fought recurrent infections that consumed their lungs and brains,[1] and their families maintained bedside vigils during prolonged hospitalizations. Although many lost husbands, sons, and brothers, they continued their strong support for better health care and more research.

Early in the epidemic, people feared that merely being in the vicinity of an AIDS-infected person was dangerous because the modes of transmission of the virus were unknown. Some physicians urged that patients with AIDS should be removed from hospitals because caring for them was "too hard, too harsh, and too hopeless" [8]. But many doctors, nurses, medical technicians, dentists, and dental hygienists continued to provide essential services to their patients despite being aware of the risks of exposure to HIV-contaminated blood and the possibility of needle-stick injuries. Although there were some surgeons who refused to operate on HIV-infected persons, many others performed essential operative procedures despite knowing that a slip of the scalpel could expose their bodies to infected blood. These courageous individuals functioned in the best tradition of the health care profession and are the unsung heroes of the HIV epidemic.

---

1. *Pneumocystis jirovecii* pneumonia and progressive multifocal leukoencephalopathy.

Infectious disease specialists also played a major role in the diagnosis and treatment of HIV infection. They had seen the rapid spread of AIDS among homosexuals and the often fatal outcome of the disease, and they warned that infected hemophiliacs would likely suffer the same fate. They educated patients and their spouses about the risks of sexual transmission of HIV and provided expert management for the many exotic infections that affected people with AIDS. Without the dedication of these and other professionals, the outcomes of the epidemic would have been far worse, and they are recognized here for their advice and support.

KEY POINTS

- The HIV epidemic was brought under control by three key breakthroughs: the development of a test for the virus; the availability of sterilized clotting factor concentrates; and the discovery of effective antiretroviral medications.
- A few exemplary individuals recognized the danger that AIDS posed to the hemophilia community. They warned about the risks of clotting factor concentrates and provided exemplary leadership during the epidemic.
- Many hemophiliacs and their families advocated for better health care and more support for AIDS research and education.
- Physicians, dentists, and other health care personnel continued to provide care for their patients despite the risk of infection from exposure to HIV-contaminated bodily fluids.

REFERENCES

[1] Saxinger WC, Gallo RC. Application of the indirect enzyme-linked immunosorbent assay microtest to the detection and surveillance of human T cell leukemia-lymphoma virus. Lab Invest 1983;49:371–7.
[2] Safai B, Sarngadharan MG, Groopman JE, et al. Seroepidemiological studies of human T-lymphotropic retrovirus type III in acquired immunodeficiency syndrome. Lancet 1984;i:1438–40.
[3] U.S. Department of Health and Human Services/Public Health Service. Update: acquired immunodeficiency syndrome (AIDS) in persons with hemophilia. MMWR Morb Mortal Wkly Rep 1984;33:589–91.
[4] Ryan J. Letter from Cutter Laboratories; October 26, 1984.
[5] Chaisson RE, Allain J-P, Volberding PA. Significant changes in HIV antigen level in the serum of patients treated with azidothymidine. N Engl J Med 1986;315:1610–11.
[6] Volberding PA, Lagakos SW, Koch MA, et al. Zidovudine in asymptomatic human immunodeficiency virus infection. A controlled trial in persons with fewer than 500 CD4-positive cells per cubic millimeter. The AIDS Clinical Trials Group of the National Institute of Allergy and Infectious Diseases. N Engl J Med 1990;322:941–9.
[7] Shilts R. And the band played on. New York, NY: St Martin's Press; 1987. pp. 115–16.
[8] Cohen RL. Life, death, and AIDS. Ann Intern Med 1999;131:158.

Chapter 11

Hemophilia: Past and Present

The decade of the 1980s was a horrendous time for hemophiliacs and their families. The epidemic was a tragedy on many levels. First, nearly 40% of them died of acquired immunodeficiency syndrome (AIDS); because they feared getting AIDS from blood products, hemophiliacs reduced their dose or completely stopped infusing clotting factor concentrates. As a consequence, they experienced a resumption of painful hemorrhages and developed progressive disability. Second, even if a hemophiliac was able to attend school or work, he was often denied school attendance and access to his workplace. Third, conjugal relationships were strained because of concerns about transmission of the virus to spouses by unprotected sex, and starting a family had to be postponed indefinitely because of the risk of infecting the mother and unborn child. Finally, boys and men who had learned to cope with hemophilia were now at risk for developing AIDS and could experience fever, weight loss, swollen glands, diarrhea, and often fatal infections such as pneumonia and encephalitis. Many young lives were lost and families were decimated.

HEMOPHILIA CARE AT THE TIME OF THE HIV EPIDEMIC

In the early years of the epidemic, the relationship between hemophiliacs and their physicians was severely tested. People seeking medical attention expect that their doctors will be able to accurately diagnose their illness and prescribe the appropriate treatment. Although physicians learn about some diseases during their training, the bulk of their knowledge usually comes when they open a practice and start seeing sick patients. That is when they discover that most diseases have not "read the textbook" description of their classic appearance. Disease features are enormously variable because disease-causing microbes have evolved to evade host defenses and seek out the most vulnerable tissues to attack; this might be the respiratory tract in one person or the gastrointestinal system in another. Thus, unpredictability in disease presentation often accounts for the misdiagnosis of even common disorders, let alone conditions physicians have not previously encountered and cannot readily reference, such as hemophilia and AIDS. Consequently, the care provided by the medical establishment was often suboptimal.

Linked by Blood: Hemophilia and AIDS. DOI: http://dx.doi.org/10.1016/B978-0-12-805302-7.00011-2

Generally, patients recognize that doctors, hospitals, and pharmacies are attempting to help them. But the earliest memories of children with hemophilia were of visits to community hospital emergency departments, where long waits under harsh fluorescent lights were common. Eventually the boys would be examined by harried physicians who usually exacerbated the pain of a joint hemorrhage by pulling and prodding the tender, sensitive tissues. Many of these doctors had never seen or treated a hemophiliac, and they were often uncertain about how to proceed, even though the patient's family members suggested giving clotting factor.[1] So, in the face of uncertainty, the easiest course for the neophyte physician was to remove the crying child from the immediate vicinity by sending him for X-rays of the painful site, thereby giving the appearance that something was being done. In the X-ray suite, the child was subject to further pain as the arm or leg was moved into various positions and held tightly while the X-ray was in progress. In reality, there was no point in getting X-rays; if bleeding was occurring into a joint for the first time, then the small amount of blood that was accumulating under the joint lining and causing pain would not be visible on the film, and if there had been repeated hemorrhages into this joint, then the X-ray would show evidence of previous tissue destruction. But neither of the X-ray findings would alter the immediate need for treatment to control the new bleeding.

Upon the patient's return from the radiologic suite, the emergency department doctor would call the patient's physician or senior staff member, who almost always concurred with the family that an infusion of clotting factor should be given. At this point, the doctor would request the medication from the hospital pharmacy or blood bank. If the patient was lucky, the clotting factor would be available from these sources; if the patient was unlucky (as almost always was the case when the emergency department was located in a community hospital), the doctor would be informed that the hospital did not stock clotting factors and the patient would be directed to a facility that could provide the medication. Usually, this was a major medical center located many miles from the community hospital, requiring the patient and his family to travel long distances, often in the middle of the night. This delay in giving definitive therapy allowed bleeding to continue, escalating pain and destroying more tissue.

When the hemophiliac and his family arrived at the larger medical center, they again were required to complete the registration process and wait to be seen by the triage nurse and the emergency department physician. Another physical examination was performed, a staff physician was contacted, and the coagulation products were ordered. Because clotting factors are complex proteins, they need

---

1. For example, Jean White, Ryan White's mother, recalled an incident when Ryan lacerated his head. She brought him to the emergency department and requested that he receive clotting factor, but the doctor assured her that just bandaging the laceration would suffice. They left the hospital, but later that night there was profuse bleeding from the wound. Mrs White realized that she, not the doctors, would have to make the decisions about Ryan's care. She concluded that she had more knowledge about hemophilia than they did.

to be injected directly into the blood stream to be effective. If the hemophiliac was a young child, placing and maintaining a needle in a small vein was a daunting task. The screaming boy would be physically restrained by one or two staff members while the doctor sought to access the vein; multiple punctures were usually required before there was satisfactory placement of the needle. Then, the medicine would be infused and the exhausted child would fall asleep. Although many of these difficulties were resolved by the introduction of home treatment in the 1970s, the memories of these traumatic events were imprinted indelibly on the consciousness of many hemophiliacs treated before this era and have contributed to hostility toward the clumsy and inefficient health care system.[2]

There were other reasons for animosity toward doctors and their assistants. Boys wanted to be active in sports, ride bikes, roughhouse, and engage in a variety of other pursuits that could result in bleeding. When parents asked whether their sons could participate in such activities, most physicians responded negatively. Whenever a boy wanted to play, the parents told him: "Doctor says you are not allowed to do that." This reinforced the negative attitudes toward physicians. As the child grew older, he recognized that his disease was expensive and that the costs of his care were depriving the family of opportunities to move to finer houses, acquire luxury items, and take vacations. Economic hardships presented to the family by health care expenses contributed to resentment of the medical establishment.

When the cause of AIDS in hemophiliacs was linked to the blood products that had been prescribed for them, many patients saw vindication for their hostile attitudes toward physicians. They had trusted their care to doctors and now felt betrayed. After all, human immunodeficiency virus (HIV) was an iatrogenic disease; doctors had recommended the products that transmitted the infection. They were also furious with the National Hemophilia Foundation (NHF) and its Medical Advisory Board. Many resigned their memberships and started their own organizations, conspicuously avoiding inclusion of any medical personnel. Several hemophiliacs and their families instituted lawsuits against physicians, pharmaceutical companies, and others. Doctor–patient relationships deteriorated as the epidemic surged through the 1980s and into the 1990s, but they improved with the advent of safer hemophilia medications and effective treatment for HIV infection.

HEMOPHILIA CARE SINCE THE HIV EPIDEMIC

The AIDS epidemic in hemophiliacs was finally brought to a close by procedures that inactivated the virus. The products currently prepared from human plasma are quite safe with respect to transmission of HIV and most hepatitis

---

2. Visits to the emergency department (sometimes called the emergency room (ER)) are still problematic for people with hemophilia and their families. The Hemophilia Federation of America has issued a pamphlet entitled "Surviving the ER" that provides advice about what to do before an emergency, just before leaving for the emergency department, and when arriving at the emergency department.

viruses. But there are certain viruses, such as parvovirus B19 and hepatitis A virus, that can slip through the viral inactivation steps included in the plasma purification process. There is also concern about contamination with the agents associated with Creutzfeldt–Jakob disease (CJD) and variant CJD [1]. These are tiny infectious particles called prions, and they have caused severe neurologic disease in cattle and humans. Although studies have not identified these agents in brain tissue from persons with hemophilia [2], they are readily transmitted by blood transfusion to sheep [3]. Thus, the safety of products derived from human blood can never be completely assured.

Concerns about the possibility that plasma-derived concentrates could be contaminated by infectious agents led to the development of products prepared by recombinant DNA technology [4]. The first step in this process required the extensive purification of clotting factors VIII and IX, followed by the elucidation of their molecular structures. Next, the genes for these factors were cloned by workers at the Genetics Institute in Boston and at Genentech in South San Francisco, and the recombinant proteins were prepared for clinical use. Dr Gilbert White and colleagues gave the first infusion of recombinant factor VIII to two hemophilic patients at the Harold R. Roberts Comprehensive Hemophilia Diagnosis and Treatment Center at the University of North Carolina and showed that the recombinant protein was safe and effective for controlling bleeding [5]. After extensive clinical trials, the recombinant clotting factors were licensed by the FDA; currently, only recombinant products are recommended for the treatment of hemophilia. The NHF urges manufacturers to avoid using human or animal proteins in these biological agents and to keep their prices in line with those of plasma-derived clotting factor concentrates.

Comprehensive care of hemophilia is not limited to simply providing blood products. Medications that safely relieve the pain of joint hemorrhages are also important, but only drugs that do not contribute to the bleeding tendency are suitable. Aspirin is not recommended because it can cause stomach irritation and interferes with normal platelet function. Medications to relieve pain may be given orally or intravenously, but injections into the muscle are avoided because of the danger of provoking a hemorrhage. However, vaccines that are injected under the skin are without hazard; all hemophiliacs are urged to undergo the usual childhood vaccinations as well as vaccinations for hepatitis A and hepatitis B. Another major component of comprehensive care is education; the rarity of hemophilia means that not many health care providers are familiar with its clinical features. These caregivers might not recognize that bouts of severe pain could be due to bleeding into a joint, and they might assume that large bruises are signs of child abuse rather than a hemorrhagic disorder. Other persons needing education are family members who often have little knowledge about the unique mode of inheritance of hemophilia; they might not know that daughters of hemophiliacs always carry the trait and can transmit it to their offspring, whereas the sons of hemophiliacs never pass the disorder to subsequent generations. Finally, psychosocial support and financial counseling are essential for patients and families; to obtain these, the services of a Hemophilia Comprehensive Care Center should be sought as soon as the disease is diagnosed.

Hemophilia is a global disorder that affects all races and societies. The World Hemophilia Foundation (WHF) was organized in 1963 and has affiliates in 122 countries. Its mission is to train caregivers in the diagnosis and treatment of bleeding disorders, to advocate for an adequate supply of safe therapeutic agents, and to educate and empower its constituents and the public. However, it has been difficult for the WHF to achieve these goals in many parts of the world; hemophiliacs have inadequate treatment in as many as 75% of the countries surveyed by the WHF [6]. If bleeding is uncontrolled because clotting factor concentrates are unavailable, then hemophiliacs can lose large amounts of blood and require blood transfusions. This puts them at risk for HIV infection because some of the blood donors in poorer countries are HIV-positive. It is ironic that hemophiliacs in the third world avoided infection by contaminated commercial concentrates (they couldn't afford them) but nevertheless became infected by blood containing HIV.

In the three decades since the advent of the AIDS epidemic, new HIV infections in US hemophiliacs have been extremely rare, but HIV persists in those infected during the early 1980s. As of March 2015, 18.4% of US hemophiliacs between the ages of 30 and 74 were positive for HIV [7]. Fortunately, treatment for HIV infection has progressively improved, and HIV-related deaths in hemophiliacs are now rare. In addition, attitudes toward HIV-infected individuals have gradually grown more tolerant. This change in attitude occurred when people learned that they were not at risk from casual contact with HIV-infected persons. In addition, tolerance for those with differing sexual orientations has grown. However, hemophiliacs in the rest of the world have not fared as well.

Worldwide, many HIV-positive people still experience the same discrimination that characterized the popular response to this disease in the United States in the 1980s. As many as 78 countries in the developing world have enacted laws that severely punish homosexual behavior, and some officials equate AIDS with homosexuality. The consequence of such laws is that HIV-infected individuals, even if the infection is not sexually related, might be reluctant to seek medical care for fear that they will be branded as homosexuals. Undiagnosed HIV-infected people unwittingly transmit the disease to their sexual contacts as well as to the recipients of their donated blood. There are currently 37 million HIV-infected people eligible for antiretroviral treatment under the new World Health Organization (WHO) guidelines [8], and an estimated 2 million new infections and 1 million deaths each year. One could argue that hemophiliacs in developing countries are worse off now than in the 1970s because they still might not have access to clotting factor concentrates, are at risk for HIV from blood transfusions, and are ostracized by their communities if they become HIV-positive.

Discrimination and bigotry toward persons with AIDS is not limited to third-world countries. Recently, Los Angeles Clippers basketball team owner Donald Sterling said that Magic Johnson, the former basketball star, should be ashamed of himself for contracting HIV [9]. The *New York Times* columnist Charles M. Blow describes this attitude as AIDS shaming, because it suggests that HIV-infected persons have a character defect. As Blow points out, HIV is a communicable disease and should be treated as such. The stigma attached to AIDS, cancer, and other diseases is completely inappropriate, and hopefully a better-educated public and effective therapy will end these prejudicial attitudes.

Although replacement of the missing or defective clotting factor controls or prevents bleeding, hemophiliacs must still infuse the concentrates intravenously as often as two or three times per week. Because this treatment is burdensome and expensive, people with hemophilia are seeking a cure for their disease. The approach currently being studied in clinical trials is gene therapy. Just as a surgeon can repair a congenital cleft palate, so might a gene therapist operate on a defective clotting factor gene and remove and replace the mutated segment. This feat can be accomplished by using a virus to insert a healthy gene into the cell containing the defective gene. Once the normal gene is introduced into the cell, it should be possible for that cell to make a fully functional clotting factor. Although seemingly straightforward, this technique has encountered several major difficulties. For example, the recipient might view the viral carrier as an invader and destroy it, the gene might be inserted into an area of the DNA that increases its susceptibility to becoming cancerous, and the newly formed clotting factor might be considered "foreign" by the immune system of the recipient, provoking an immunologic reaction resulting in its inactivation. These problems are gradually being overcome by enlisting more refined viral vectors, using targeted genome editing and designer enzymes [10], and using clotting factors modified to decrease their immunogenicity. Recently, it was announced that two pharmaceutical companies have entered into a collaboration to develop and commercialize a novel gene therapy for the treatment of factor VIII deficiency hemophilia [11], and two other companies are sponsoring a clinical trial for those with factor IX deficiency [12]. Gene therapy has the potential to free hemophiliacs from their dependence on replacement blood products and the risk of future blood-borne infections, as well as the crippling financial toll of this disease.

ACCESS TO PHARMACEUTICAL PRODUCTS

The development of effective clotting factor concentrates has dramatically improved the outlook for hemophilia, but the cost of these therapeutic materials remains a problem for many individuals. Hemophilia has become a multibillion-dollar business opportunity for the small number of suppliers of clotting factor concentrates in the United States. As noted in chapter 1, page 7, the price of these therapies has dramatically increased as manufacturers implemented more stringent purification procedures and replaced plasma-derived products with recombinants. The recombinant concentrates are regularly modified to achieve small improvements in potency, purity, and duration of action, with a concomitant increase in the price of each "new" product. These modifications might fall into the category of "product hopping," a tactic used by drug makers to extend the patent life of their products and blunt competition from generic versions [13]. The companies vigorously defend their patents, ensuring that there are no competing generic products. Manufacturers claim that they need to raise prices to recoup money spent on research but, in fact, they spend large sums on promoting their drugs to doctors and in direct-to-consumer advertising.

Persons with hemophilia must perform a cost-benefit analysis for each new high-priced concentrate, weighing symptoms, disposable income, and insurance coverage. For example, Eloctate (Biogen Idec Inc.) is a new, longer-acting factor VIII concentrate. Its advantage over current products is that dosing to prevent bleeding is required only twice rather than three times per week. The reduced frequency of intravenous injections is an advantage, and one might think that fewer doses would reduce treatment costs, but Eloctate is priced so that it will cost as much as most currently available concentrates. The new products and many older concentrates are too expensive for most hemophiliacs in developing countries who suffer recurrent hemorrhages and become crippled because they cannot afford the high cost of the clotting factors.

Currently, severe hemophiliacs have more than $100,000 in annual expenses for therapeutic materials, including pain medications and drugs for hepatitis and HIV infections. Prices for these medicines have escalated in recent years, with spending on prescription drugs in the United States increasing 13% in 2014 alone [14]. A few extremely pricey agents have been the main contributors to this expense; the table shows the sales in billions of dollars for some of the new drugs for hepatitis B/C and HIV/AIDS [15]. The cost of one drug for the treatment of hepatitis C infection is $84,000 for a standard 12-week course of therapy, and sales of this antiviral drug totaled $2.3 billion in the first quarter of 2014 alone [16]. Pharmaceutical companies are entitled to reasonable returns on investments in technology and facilities, but a net income of $2.3 billion in 3 months seems excessive!

Drug	Disease	Sales in billions ($)
Lamivudine	Hepatitis B	0.2
Sofosbuvir	Hepatitis C	10.3
Combivir	HIV/AIDS	0.1
Truvada	HIV/AIDS	3.3

Worldwide, these new medications are too expensive for most of the 185 million persons with hepatitis C infection [17,18], and high drug costs probably explain why fewer than 14 million of the estimated 37 million HIV-infected people are using antiviral agents [19].

Hepatitis C is not a rare disease; those infected include hemophiliacs who were exposed to the virus when they received treatment with unpasteurized clotting factor concentrates, army veterans who contracted the infection when stationed abroad, and as many as 17% of persons in our prison system [20]. It is estimated that much of the recent increase in the cost of medications is due to the hepatitis drugs. The Veterans Health Administration notes that the new hepatitis C treatment "has greatly outpaced V.A.'s ability to internally provide all aspects of this care" [21]. In addition, states are obligated to provide medical care for their prisoners; the cost for treating all those incarcerated could be as high as $33 billion, threatening the financial stability of several states [22]. The consequences of not providing the drugs are costly hospitalizations due to disease complications, such as cirrhosis and liver failure, and the spread of the virus by infected persons sharing contaminated needles with previously uninfected drug abusers.

To obtain the hepatitis drugs at a lower cost, governments might try to force pharmaceutical companies to permit the manufacture of generic substitutes, as has been done for HIV drugs. The actual cost to manufacture the new hepatitis C drugs is estimated to be no more than $78–166 per person for a 12-week course [11], an expense that would be economically bearable for most governments. Possibly because of public outcry over the huge profit from Sovaldi (the drug would be profitable even with a 99% cut in its price), Gilead Sciences has agreed to allow the sale of lower-cost generic versions in India and some other developing countries [23].

The current system of patent protection has enabled pharmaceutical companies to price their products at levels greatly in excess of the cost of manufacture, without fear that competitors will challenge their market share. They use practices such as "product hopping" to ensure that competitors with lower-priced generics will not be able to enter the market. Even after the patent has expired, manufacturers might tightly control drug distribution so that generic companies will have difficulty obtaining the samples they need for required testing. Generic drugs are also becoming more expensive; the costs for essential medicines made by only a few generic companies have more than doubled in the past year [24]. Another tactic is to acquire companies making older drugs and rebrand these medications as high-priced "specialty drugs" [25a,b]. Although costs for some increased modestly (naloxone, $14.90–34.50 per tablet [26]), others have increased dramatically (Daraprim, $13.50–750 per tablet [24]). People on long-term treatment with Colchicine, Isuprel, Nitropress, or Cycloserine suddenly saw a huge escalation in their pharmacy charges; annual costs now exceed $200,000 for some drugs.[3] The often exorbitant drug prices benefit shareholders at the expense of those who desperately need these medications.

---

3. Acthar Gel, $205,681; Cinryze, $230,826; Kalydeco, $299,592; Naglazyme, $485,747; Soliris, $536,629 (FiercePharma, "The Top 10 Most Expensive Drugs of 2013").

A champion of making essential drugs more affordable is Hagop M. Kantarjian, MD, an oncologist at the University of Texas MD Anderson Cancer Center. He has been an outspoken proponent of a patient-driven petition to combat the high costs of drugs [27]. The Petition is addressed to the US Secretary of Health and Human Services, the Congress, and the President, and includes the following recommendations [28]:

- Allow Medicare to negotiate drug prices with pharmaceutical manufacturers and the new Patient-Centered Outcomes Research Institute to include drug prices in their assessments of drugs and treatments
- Enact new legislation that prevents drug companies from delaying access to generic drugs (Pay for Delay) and extending the life of drug patents (Patent Evergreening)
- Create a mechanism for patients and their advocates to propose a fair price for new treatments based on value to patients and health care.

Allowing access to life-saving medicines for the global community is essential for world health. If patients infected with hepatitis C, HIV, or other viruses such as Ebola remain untreated, then these diseases will spread throughout the local population and eventually to people in other parts of the world. It is to everyone's benefit to ensure that persons with these infections receive curative therapy; to make these drugs accessible, there should be government–industry partnerships. Governments need to recognize that manufacturers are entitled to be recompensed for their costs and rewarded, within reasonable limits, for bringing new drugs to market. If pharmaceutical manufacturers do not voluntarily reduce their prices, then there are a few alternatives. They might be pressured to allow generic companies to mass-produce drugs for low-income countries at a cost close to the cost of production, with a small royalty returned to the pharmaceutical companies; alternatively, countries can overrule company patents on drugs and import generics at lower costs. These are issues that are currently being addressed by the international community and hopefully will bring about better treatment for all diseases.

A proposal for the control of prices of generic and brand name drugs in the United States is described in the next chapter.

INFECTIOUS DISEASE EPIDEMICS

It seems that we are confronted by new challenges to our health on almost a daily basis; the most recent examples are the 2014 Ebola epidemic affecting several West African nations and the spread of Zika virus throughout Central and South America. There are many parallels between AIDS and Ebola viral disease (EVD): both infections were first reported from Zaire in Central Africa—AIDS in 1960 and EVD in 1976; both are spread by direct contact with bodily fluids; and both result in a severe, often fatal, illness. Curative antiviral agents and an effective vaccine are not yet available for either infection. Just as fear and panic gripped communities when AIDS first surfaced, similar irrational emotional responses to EVD are reported, even though there is no risk from indirect contact, such as being on the same airplane with an infected person [29]. Survivors of EVD are no longer contagious but have been shunned by family, friends, and neighbors [30]. Children have been barred from schools and adults have been prevented from entering their workplaces, resulting in severe economic hardship and mental anguish. Ignorance about the origins of these diseases has spawned conspiracy theories; for example, that the AIDS virus was unleashed to destroy the homosexual community and that white bureaucrats are using Ebola to depopulate Africa. These irrational concerns and attitudes can be addressed by intensive educational campaigns using the experience previously gained from AIDS and hemophilia as a guide. Stigmatizing infected individuals is inappropriate; becoming better informed about disease benefits everyone. Ignorance can be dispelled by infectious disease experts, who can use electronic media to communicate the most current information to community leaders.

In our global one-world community, infectious agents do not remain confined to one locale, but rather have the potential to become broadly dispersed. Just as the AIDS epidemic was restrained by implementing measures for safe sex and closing bathhouses, so will EVD be curbed by limiting the spread of the disease in West Africa. Therefore, we should support agencies, such as WHO and the Centers for Disease Control (CDC), that engage in the control and eradication of disease. Furthermore, the experience of HIV infection in hemophilia shows that we must be continually vigilant about the safety of the blood supply and screen potential donors for a broad array of infectious agents. Testing for these microbes will require investments in techniques for viral isolation, characterization, and detection. Efforts must also be expended on the discovery of new therapeutic materials and vaccine development. As Gilbert White notes, the importance of research in leading to scientific advances in hemophilia cannot be overemphasized [4]. Research is foundational for disease treatment and prevention, but exciting discoveries must be verified by independent sources and new drugs must be tested by clinical trials. Most importantly, new therapeutic agents should be made accessible to all who need them, and this might require giving greater authority to regulatory agencies. Finally, we must all be advocates for those burdened by disease so that they can achieve their maximum potential in society.

During the 1970s, most hemophiliacs began to use clotting factor concentrates because these products dramatically improved their lives and enabled them to enter the mainstream of society. They and their physicians considered these therapeutic materials to be indispensable for the modern treatment of hemophilia. The decade of the 1970s also saw the emergence of the gay liberation movement and the appearance of bathhouses and sex clubs that allowed homosexuals to engage in unfettered sexual activity. One HIV-infected individual could spread the virus to hundreds of partners. Homosexuals were a major source of the plasma needed to prepare clotting factor concentrates; during the

1970s, they donated 5–9% of the blood collected by Irwin Memorial Blood Bank in San Francisco [31]. Because concentrate manufacture required the use of plasma from thousands of donors, the inclusion of a single plasma donation from an infected person could contaminate an entire lot of concentrate. Potentially infected material was used for the treatment of hemophiliacs in the United States and shipped abroad. The worldwide decimation of the hemophilia community was a consequence of the convergence of hemophilia and AIDS in the 1970s, linked by blood.

KEY POINTS

- Between 1981 and 1985, hemophiliacs were tremendously stressed: their choices were to treat hemorrhages and risk contracting AIDS or withhold therapy and suffer persistent bleeding.
- Confidence in their physicians and patient organization (the NHF) was severely tested.
- The development of safe clotting factor concentrates and the introduction of effective treatment for HIV infection helped restore the faith of hemophiliacs in the medical establishment.
- In addition, public education about HIV and its transmission reduced hostile attitudes toward homosexuals and other persons infected by the virus.
- Despite these achievements, hemophiliacs in developing countries often lack access to the newer clotting factor concentrates, continue to be exposed to unsafe blood products, and are ostracized by the larger society.
- The cost of treatment for drugs effective against hepatitis viruses and HIV remains out of reach for many people, and needs to be addressed by legislative initiatives.
- Investment in education and research is vital to protect populations from the global threat of new infectious diseases.

REFERENCES

[1] Kasper CK. New variant CJD. Hemophilia Bull March 2001.
[2] Evatt B, Austin H, Barnhart E, et al. Surveillance for Creutzfeldt-Jakob disease among persons with hemophilia. Transfusion 1998;38:817–20.
[3] Houston F, McCutcheon S, Goldmann W, et al. Prion diseases are efficiently transmitted by blood transfusion in sheep. Blood 2008;112:4739–45.
[4] White II GC. Hemophilia: an amazing 35-year journey from the depths of HIV to the threshold of cure. Trans Am Clin Climatol Assoc 2010;121:61–73.
[5] White II GC, McMillan CW, Kingdon HS, et al. Use of recombinant antihemophilic factor in the treatment of two patients with classic hemophilia. N Engl J Med 1989;320:166–70.
[6] <www.WHF.org>; [accessed 22.05.14].
[7] American Thrombosis and Hemostasis Network (ATHN) dataset; March 31, 2015.
[8] Anonymous. By the numbers. News in Brief, Science 2015;350:142.
[9] Blow CM. The AIDS-shaming of Magic Johnson. NY Times 2014;May 14.

[10] Cathomen T, Ehl S. Translating the genomic revolution-targeted genome editing in primates. N Engl J Med 2014;370:2342–5.

[11] Bayer news release dated June 23, 2014. Quoted in Handi, a publication of the National Hemophilia Foundation. Bayer and Dimension partner to develop novel hemophilia A gene therapy; June 30, 2014.

[12] Gene-therapy, open label, dose-escalation study of SPK-9001 in subjects with hemophilia B (NCT02484092), initiated December 2014.

[13] Federal Trade Commission. FTC files amicus brief explaining that pharmaceutical "product hopping" can be the basis for an antitrust suit; November 17, 2012.

[14] Kapczynski A. The trans-pacific partnership—is it bad for your health? N Engl J Med 2015;373:201–3.

[15] Modified from Cohen J. King of the pills. Science 2015;348:622–5.

[16] Pollack A. Gilead revenue soars on hepatitis C drug. NY Times 2014;April 23.

[17] Hill A, Cooke G. Hepatitis C can be cured globally, but at what cost? Science 2014;345:141–2.

[18] Editorial. How much should hepatitis C treatment cost? NY Times 2014;March 16.

[19] Editorial. Treating HIV patients before they get sick. NY Times 2015;May 31.

[20] Rich JD, Allen SA, Williams BA. Responding to hepatitis C through the criminal justice system. N Engl J Med 2014;370:1871–4.

[21] Quoted by Oppel Jr. RA. Wait lists grow as many more veterans seek care and funding falls far short. NY Times 2015;June 21:12.

[22] Sanger-Katz M. Boon for hepatitis C patients, disaster for prison budgets. NY Times 2014;August 7:A3.

[23] Harris G. Deal struck on generics of costly hepatitis C drug. NY Times 2014;September 16:B1.

[24] Rosenthal E. Rapid price increases for some generic drugs catch users by surprise. NY Times 2014;July 9:A16.

[25a] Pollack A. Once a neglected treatment, now an expensive specialty drug. NY Times 2015;September 21:B1–B2.

[25b] Pollack A, Tavernise S. A drug company's price tactics pinch insurers and consumers. NY Times 2015;October 5 p. A1, B2.

[26] Goodman JD. Attorney General critical of heroin antidote's cost. NY Times 2014;December 2:A21.

[27] Lawrence L. Efforts to protest high cancer drug prices underway. ASH Clin News May 2015.

[28] The petition is entitled "Protest High Cancer Drug Prices so All Patients With Cancer Have Access to Affordable Drugs to Save Their Lives" and is on the website <www.Change.org>.

[29] Lipsey S. Just plane scared: can you get Ebola from an airplane? Accessed from: <www.yahoo.com>; October 16, 2014.

[30] Nossiter A. Surviving Ebola, but untouchable back home. NY Times 2014;August 19: pp. 1, 8.

[31] Shilts R. And the band played on. New York, NY: St Martin's Press; 1987, p. 199.

Chapter 12

A Prescription for the Next Health Care Crisis

BLOOD PRODUCT SAFETY

As long ago as 1975, the World Health Organization (WHO) promoted the development of national blood services based on voluntary non-remunerated donation of blood [1]. This action was prompted by a report that posttransfusion hepatitis was more likely if the blood came from paid donors [2]. Although blood banks switched to recruiting only volunteer donors, HIV-infected blood was still able to enter the blood collection and distribution system. In the wake of the acquired immunodeficiency syndrome (AIDS) epidemic, it became clear that additional measures were needed to prevent microbes from entering the blood supply, and blood banks adopted very extensive screening of donors and testing of donated blood. Potential donors now complete questionnaires and are deferred, or the blood is not used for transfusion if donors fall into any of the categories listed in chapter 3, page 33. Blood banks also increased the number of infectious agents included in their testing panels; the tests of donor blood currently required or recommended are likewise displayed in chapter 3, page 33. These efforts have been remarkably successful in preventing contamination of the blood supply, and the risk of contracting AIDS from a transfusion in the United States is estimated at less than 1 in every 2 million units transfused [3].

The Food and Drug Administration (FDA) is the main government agency tasked with safeguarding the nation's blood supply. It was established by the 1906 Pure Food and Drug Act and is a vast agency with four directorates overseeing medical products and tobacco, foods and veterinary medicine, global regulatory operations and policy, and operations. The principal regulatory agency with responsibility for the safety of blood and blood products is the Center for Biologics Evaluation and Research (CBER) within the Office of Medical Products and Tobacco. Its Blood Products Advisory Committee reviews and evaluates data concerning the safety, effectiveness, and appropriate use of products derived from blood or biotechnology intended for use in the diagnosis, prevention, or treatment of human diseases. CBER regulates the collection of blood and blood components and establishes standards for the products themselves.

Linked by Blood: Hemophilia and AIDS. DOI: http://dx.doi.org/10.1016/B978-0-12-805302-7.00012-4

It evaluates scientific and clinical data submitted by manufacturers to determine whether products meet standards for approval, and it determines whether they have reasonable risks given the magnitude of the benefit expected and the alternatives available. In addition, it inspects blood establishments and monitors reports of errors, accidents, and adverse clinical events. It is responsible for identifying and responding to potential threats to blood safety, to develop safety and technical standards, and to help industry promote an adequate supply of blood and blood products. The agency is charged with providing up-to-date information to the public, health care professionals, the media, and product manufacturers through its Biologics web pages and Patient Network Newsletter.

The Institute of Medicine (IOM) in its 1995 report presented an analysis of the FDA and its Directorates, and concluded that there needed to be a "far more responsive and integrated process to ensure blood safety" [4]. They recommended the creation of a "Blood Safety Council" to assess and propose strategies for overcoming current and potential future threats to the blood supply and to educate public health officials, clinicians, and the public about challenges to blood safety and strategies for dealing with these challenges. They suggested that the Council could also alert scientists about the needs and opportunities for research to maximize the safety of blood and blood products. These IOM proposals were never implemented, but given the recent epidemics of previously exotic contagious diseases such as Ebola and Dengue hemorrhagic fever, and the emergence of SARS and MERS (severe acute respiratory syndrome and Middle Eastern respiratory syndrome, respectively), establishing an office to safeguard the blood supply would seem to be even more urgent now than when they proposed it in 1995. Blood banks do not currently test donors for these viruses, so there is the possibility that these organisms could be transmitted by transfusion. An even greater risk to the blood supply of the United States is the Chikungunya virus, which appeared in the Caribbean basin in December 2013. This virus causes fever and joint pains and has already infected approximately 15,000 persons. A group of French investigators screened plasma samples from 2149 blood donors from Martinique to find out if this virus is capable of gaining entrance to the blood supply [5]. The Chikungunya virus was found in the blood of four men; two never had symptoms of infection and the other two had fevers that began soon after their donations. Had their blood been used for the preparation of clotting factor concentrates or transfused into patients, it could have transmitted the virus to the recipients. This experience clearly shows that our blood supply remains vulnerable and requires continuous surveillance for new sources of contamination.[1]

Because people with hemophilia are major consumers of blood products, the Medical and Scientific Advisory Council (MASAC) of the National Hemophilia

---

1. In 2016, the W.H.O. issued an alert that the rapid spread of Zika virus posed a threat to the blood supply (Louis CS. W.H.O. issues guidance on blood in Zika areas. The NY Times, February 20, 2016, p. A5.

Foundation (NHF) has submitted to the FDA a number of recommendations for improving the quality of the plasma used for preparing clotting factor concentrates [6]. Among these are suggestions that the plasma pool size should be limited to 15,000 donors, very sensitive tests should be used to detect viral contamination, and improved techniques should be implemented for eliminating infectious agents. In addition, NHF requests that manufacturers should promptly report suspected infections associated with their products; any such products should be assumed to be implicated in disease transmission and removed from the distribution path and patients' homes. Another recommendation is that the FDA should communicate promptly with consumer organizations such as NHF whenever there is a recall or voluntary withdrawal, because these actions could have an impact on the supply and availability of clotting factor concentrates. They also suggest using bar-coding to identify coagulation products; adopting this method would facilitate accurate tracking and dispensing, as well as usage in the hospital and at home. Furthermore, NHF urges that expedited regulatory review should be extended to all products offering incremental safety and efficacy advantages.

On December 2, 2014, the FDA announced that it intended to establish a general program to monitor the safety of the blood supply in collaboration with the National Heart, Lung, and Blood Institute. In addition, the FDA plans to engage in public discussions about the donation of safe blood and to review the effectiveness of the blood donor history questionnaire. Although these actions are commendable, I believe that the FDA needs to have an office specifically devoted to safeguarding blood safety.

Recommendation 1: Establish an Office of Blood Product Safety Within the FDA

This Office is needed to provide oversight for the many tasks related to the procurement, availability, and safety of blood and blood products. In addition to the functions recommended by the IOM and the NHF, the Office of Blood Product Safety (OBPS) could collect and review reports received by the FDA regarding the safety and availability of blood and blood products. The OBPS would assess the appropriateness of public dissemination of these reports; if they were deemed relevant to the health and welfare of our citizens, then OBPS personnel would interpret and translate them into lay language. This material would be posted on a CBER website specifically designed for communicating important issues to the public. Paper copies of the material would be made available for persons without computer access, and the information would also be sent to nonprofit groups such as the NHF and the American Thrombosis and Hemostasis Network. These organizations have registries of people with bleeding disorders; therefore, they could be instrumental in ensuring that the information is distributed to these individuals. This method of disseminating information about product safety would replace the current piecemeal notification systems used by the NHF and other consumer organizations.

The OBPS might also address the important issue of spotty geographic availability of clotting factor concentrates. As noted, these concentrates are vital for the well-being of persons with bleeding disorders, which is why they maintain a supply of these products in their home or office. However, there are times when hemophiliacs have hemorrhages while traveling for business or pleasure. A visit to the nearest emergency department often brings the unwelcome news that the facility does not stock clotting factor concentrates. Surprisingly, even trauma centers designated level I (most capabilities) might not have coagulation products. Such facilities must either call a distribution center and request immediate delivery of the concentrate or transfer the bleeding patient to a treatment center that has the appropriate material. The lack of immediate accessibility to clotting factor creates delays in the provision of care, and the bleeding that continues unabated exacerbates the damage in whatever organ has been the site of the hemorrhage, be it the brain, joint, or other tissue. The OBPS might mandate that adequate supplies of clotting factor concentrate must be physically present in all level I trauma centers and other facilities that serve communities where persons with hemophilia and other bleeding disorders reside.

The FDA appears to recognize that its communication with the bleeding disorders community has been suboptimal. It has announced a new Patient-Focused Drug Development Initiative whose purpose is to provide opportunities for affected individuals to inform FDA officials about the treatments they consider most valuable. The FDA believes its rule-making will be enhanced if it understands patients' tolerance for benefit/risk tradeoffs. The proposed OBPS could serve as the vital interface between the FDA and the consumers of blood products, obtaining input from persons with bleeding disorders as well as disseminating information about the safety and availability of clotting factor concentrates and other coagulation products.

CONTROLLING THE COSTS OF HEMOPHILIA THERAPY

The major improvements in blood safety in recent years have been accompanied by large increases in the price of a transfusion. There is no cost for the blood itself; donors are not financially compensated for donations given for the benefit of those who are ill. However, the recipients of blood must pay a steep price for transfusions. While the cost to hospitals to purchase a unit of blood is usually $225–240, patients might be charged $1000 or more [7]; this is often loosely justified by claiming expenses for blood administration, inventory losses, and liability insurance, although formal cost-accounting is rarely available.[2] The monies collected for blood have made blood banking a multibillion-dollar enterprise, often generating large annual surpluses [8].

---

2. The University of Utah Health Care is one of the few hospitals with a computer program to cost-account items for patient care (Kolata G. What are a hospital's costs? Utah system is trying to learn. NY Times September 8, 2015, pp. A1, A18).

A more enlightened system would have the costs of collecting, processing, and administering blood borne by the general public, and the prices for blood and blood products regulated by the government. Controlling the costs of transfusion and spreading the expense over the entire population would ease the financial burden on those who are ill and least able to afford this expense. Such a paradigm has been adopted by many other countries and should be incorporated into our health care system. Most other developed countries provide free care and supplies for people with chronic diseases, reasoning that diseases select their targets at random [9]. Also, they negotiate with drug and device makers to reduce list prices. For example, the British National Institute for Health and Care Excellence (NICE) is charged with assisting health and social care professionals by providing guidance on delivery of the best possible care based on the available evidence [10]. NICE conducts technology appraisals, publishes clinical guidelines, and performs cost-effectiveness analyses. It utilizes the quality-adjusted life year to measure health benefits; if the benefit is less than a designated threshold for the cost of a new drug or procedure, then the treatment is not recommended. Once the value of a therapy is established, NICE negotiates price with the manufacturer. The costs of drugs in the United States are higher because most other countries apply price controls and we do not; we rely on a competitive marketplace, but the fact is that many pharmaceutical products do not have competitors.

At present, health insurance plans are able to restrain drug prices by forcing pharmaceutical companies to compete regarding price when there are several medicines with similar indications [11]. Sometimes, drugs with significant advantages over their competitors are more expensive, or there are no competing products; in this situation, the health plan might decide that the products are too expensive and refuse to list them on its formulary. If patients wish to have such drugs, they are obliged to pay their full price out-of-pocket. This creates a double standard of care, with wealthy individuals and those with expensive insurance policies getting top-of-the-line products and others receiving only older, lower-quality material. Recently, in response to a complaint that people with HIV were subjected to restrictions on certain medications, an insurer agreed to limit the out-of-pocket costs for some of these agents [12], but this action was taken by only one company in one state. Two members of the US House of Representatives have introduced a bill (H.R.1600) to prevent private health insurance plans from imposing higher premiums for specialty drugs, but the fate of this legislation is uncertain.

Generic drugs are usually less costly than brand name medications, but prices can skyrocket if there are drug shortages, supply disruptions, or consolidations within the generic drug industry [13]. For the products used by hemophiliacs, insurers might demand that the few companies that make concentrates lower their prices, but this tactic will not be successful if all the manufacturers agree to hold the line on prices. Because hemophiliacs must have these drugs, and because third parties are obligated to pay for them, the prices are impervious to price escalation or "inelastic" [14].

Recommendation 2: Establish a New Office for the Control of Pharmaceutical Prices Within the Department of Health and Human Services

To make the care of hemophilia more affordable to patients and third-party payors, an office for the Control of Pharmaceutical Prices (COPP) should be established within the Department of Health and Human Services. It would be modeled after the Office of Price Administration (OPA) that was established by Executive Order at the start of World War II. The OPA was formed "to prevent price spiraling, rising costs of living, profiteering, and inflation resulting from market conditions caused by the diversion of large segments of the Nation's resources to the defense program, by interruptions to normal sources of supply, or by other influences growing out of the emergency" [15]. It included a Price Administration Committee that was charged with making findings and submitting recommendations for setting maximum prices, commissions, margins, fees, charges, and other elements of cost or price of materials or commodities. The OPA was successful in keeping consumer prices relatively stable during turbulent times.

Although we are not on a wartime footing, the situation with regard to many pharmaceutical products is similar; drugs indispensable for the well-being of citizens are available from only a few sources and at highly inflated prices. There is no marketplace or competition to determine the cost of many such drugs; their price is set by the manufacturers and health insurers. For the manufacturers, the considerations used to determine prices are return on investment and willingness of third parties (insurance, government) to accept the costs of the drugs (in other words, what the market will bear). Health insurers can assign some drugs, such as those used to treat HIV infection, to the highest payment tier for midlevel plans, requiring patients to contribute 40% of the cost of the drug [16]. Sick individuals are compelled to assume a heavy monetary burden at a time when their income is often substantially decreased by illness. There are few remedies for patients when insurance companies charge hefty premiums for vital medications. In addition, our government does not negotiate price with pharmaceutical companies; if fact, such negotiations were barred by Congress when Part D Medicare was enacted. As noted by Elizabeth Rosenthal, "we approve drugs and devices without considering cost-effectiveness, or even having a clue about price. We don't ask for estimates and then are surprised when the nation is stuck with a $2.7 trillion annual health care bill" [17].

COPP would have as its primary function the establishment of ceilings on prices for products for which there is currently little or no marketplace competition; these products would include essential commodities such as drugs, medical devices, blood, and blood components that are currently under the purview of the FDA. It would adopt methods for assessing cost-effectiveness that are somewhat similar to those used by the NICE Institute, and it would negotiate prices with the manufacturers based on the safety and effectiveness of their products. It could recommend that Medicare refuse payment if prices were not commensurate with the estimated value of the therapy. In that circumstance, COPP would recommend the next best alternatives. With regard to blood and blood components, it seems likely that the price of these essential substances will substantially increase in the future as methods are introduced to sterilize the final product. For example, two firms are developing a set of chemical compounds that will inactivate microbes contaminating blood [18]. Although these innovations should greatly enhance the safety of transfusion, the final blood product will undoubtedly be more expensive. We need an agency that will provide an independent assessment of the value of such a product and set limits on its price.

In addition, COPP should have the power to authorize subsidies for the production of certain pharmaceuticals that are absolutely necessary for people's health. Currently, there is little incentive for manufacturers to produce inexpensive drugs such as heparin and morphine, and even more mundane items like sterile saline and magnesium sulfate, resulting in periodic shortages of these indispensable materials [19]. Furthermore, pharmaceutical companies have little incentive for developing new classes of antibiotics, mainly because the monetary returns from such drugs are small in comparison to medicines used for chronic diseases such as cancer and diabetes. However, we desperately need new antimicrobials because of the emergence of bacteria resistant to the current drugs [20]. Vaccine production is another area frequently neglected by pharmaceutical companies because of a perceived lack of profitability. Margaret Chan, Director General of the WHO, criticized the drug industry for not developing an Ebola vaccine in advance of the current crisis. She said "A profit-driven industry does not invest in products for markets that cannot pay" [21]. Although the government has been encouraging antibiotic research by making funds available for drug development, and although Congress is considering legislation that would increase the levels of Medicare reimbursement for newer antibiotics [16], COPP could provide more immediate financial incentives for manufacturers to discover and produce novel antibiotics and vaccines. By controlling prices and ensuring the availability of essential health products, COPP would make a valuable contribution to global health care.

BUILD EDUCATIONAL PROGRAMS TO PROMOTE TOLERANCE AND INCREASE SUPPORT FOR RESEARCH

People become more tolerant when they are presented with factual information that is simple and readily understandable. Educational programs can be devised that clearly identify the cause of a particular disease (AIDS, Ebola), describe how it is spread, and indicate whether restrictions on movement or contacts are required. Most importantly, such information is essential for dispelling conspiracy theories that attribute evil intentions to health care workers, government, and others. Therefore, greater effort is needed to educate the public and remove the stigma associated with infectious diseases.

Recommendation 3: Educate the Public About Emerging Infectious Diseases

Health care agencies should enlist the assistance of professional spokespersons to broadcast educational messages, prepare videos, and write brochures that can be disseminated by electronic and print media to broad segments of the population. A major educational campaign might mitigate the stigma associated with these infectious diseases and overcome prejudicial attitudes. It might also encourage people infected by these viruses to seek medical assistance before their diseases progress and they infect others.

Many of the serious consequences of the HIV epidemic were due to prejudicial attitudes toward those with a different sexual orientation. The rampant spread of the AIDS virus among homosexuals was almost completely ignored by the media, and research into the causes and potential remedies was grossly underfunded by government [4]. This might have led blood bankers to underestimate the prevalence of HIV in donors because they failed to give attention to data derived from the homosexual population showing that infected persons might not have symptoms of AIDS for months to years. Likewise, medical experts advising the NHF were unaware that HIV infection had been fatal in almost half of those exposed to the virus and would be much worse than an untreated hemorrhage. Had these groups been better informed about what was happening in the homosexual community, they could have taken measures to limit the entrance of the virus into the blood supply and restricted the use of clotting factor concentrate, the product that infected the hemophilia community.

Public attitudes toward persons with AIDS, whether they were homosexuals or hemophiliacs, were often characterized by hostility and lack of acceptance. This was probably best exemplified by the experience of Ryan White, a boy with hemophilia and AIDS, who said that he became the target of vicious jokes, his school locker was vandalized, and his folders were marked "FAG" [22]. This occurred because AIDS was identified by many as a "gay" disease, and discrimination against homosexuals was rampant in the small town in Indiana where he grew up. White and his family rejected the idea that he was an innocent victim, because that implied that gays with AIDS were guilty. In fact, gay men gave blood altruistically and were unaware that they were infected by HIV when they donated. White understood that AIDS was a disease, not a way of life, and he and his family were grateful for the assistance and advice they received from the gay community [23]. White was a strong proponent of education and research, noting that he faced discrimination, fear, and panic but became accepted when students and parents understood the facts of his illness.

In the three decades since the advent of the AIDS epidemic, treatment for HIV infection has progressively improved and HIV-related deaths in hemophiliacs are now rare in the United States. But isolated outbreaks of the disease still occur in nonhemophiliacs. A recent epidemic in rural Indiana illustrates how the sharing of contaminated needles disseminates HIV among people who abuse drugs [24]. In this poor community, marked by abandoned homes and widespread unemployment, many individuals became addicted to an opioid painkiller called Opana (Endo Pharmaceuticals, Inc., Malvern, PA). Although this drug is meant to be taken orally, addicts crushed the pills in water and injected the slurry intravenously. They reused needles hundreds of times and shared needles with other drug abusers. HIV infection was first recognized in January 2014; there were 71 cases by December and 135 by April 2015 [25]. The transmission rate was estimated at 80%, meaning that infection occurred in 8 of every 10 individuals sharing needles with an infected person [26]. Other factors that might have contributed to the size of the epidemic are the misperception that only homosexuals could become infected by HIV and the general reluctance to undergo testing and treatment because of the stigma attached to the disease [27]. To overcome these barriers, Indiana officials instituted a program of needle exchange, offered free HIV testing, and provided clinic facilities for treatment. It is still unclear whether these measures will be effective in limiting this outbreak.

We are one society, and what happens to our neighbors affects us as well; for example, HIV can spread quickly from the homosexual to the heterosexual community, and from drug abusers to sexual partners. We need to be more tolerant of those who differ from us, whether those differences are religious, economic, or lifestyle-related. When we observe persons afflicted with disease, we need to support them and take active measures to relieve their suffering. This concern for the rights of those with AIDS was probably best expressed by Nelson Mandela, who vigorously fought against stigmatization and discrimination of infected persons [28]. We do not know who will be vulnerable to the next invasion by an infectious agent, but we must not let prejudice and irrational thinking prevent us from doing everything possible to identify and eradicate the contagion, no matter who is infected. Study of the AIDS epidemic in hemophiliacs is relevant because it shows that failure to adequately address diseases in minority populations inflicts a high cost on society as a whole; for example, it provides a convincing rationale for investing resources to control Ebola in West Africa before it is disseminated globally.

Recommendation 4: Greater Support for Basic and Applied Research

The final, and perhaps most important, component of this prescription is a plea for greater support for basic and applied research. We are on the cusp of major advances in the treatment of blood and bleeding disorders, and these breakthroughs will be accomplished if our support for the necessary research does not waver. There are at least three areas that exemplify current research progress: safer transfusion therapy, more effective clotting factors, and prevention of HIV infection.

With regard to transfusion therapy, we can never be absolutely certain that the blood of a donor is free of viruses or other microbes because of the evolution and spread of new infectious agents, such as Chikungunya, Ebola, and Zika viruses. Several methods for sterilizing blood and blood products have been approved or are undergoing development, but none has yet been mandated by the FDA [29]. The procedures being investigated include filtration, addition of solvents, detergents, or other substances, and exposure to ultraviolet light or other light sources. The efficacy and safety of the methods are still unclear, and all would increase the cost of the product.

The use of donor blood could be eliminated if it were possible to replenish the blood of anemic patients by growing their own red cells in tissue culture, but this would require the ability to generate sufficient cells to provide relief of symptoms. Investigators recently reported that functional red blood cells could be grown from the rare immature cells that normally circulate in the blood [30]. At present, the method is not clinically feasible because it produces only a limited number of red cells, is time-consuming, and uses a great deal of culture medium. But with continued research, it should eventually be possible to treat anemic patients by propagating their own cultured red blood cells and avoiding the use of other people's blood and the risk of blood-borne infection.

Another approach is to replenish hemoglobin, the oxygen-carrying protein found in the red blood cells. Such a product could be life-saving in situations in which there is massive bleeding and blood either is unavailable or cannot be typed and cross-matched in time to prevent death. Three hemoglobin products were extensively studied for this indication, but all had serious adverse effects and none received regulatory approval, although a related product has been approved for veterinary use. The current status of the various methods to limit or avoid transfusion of banked blood has recently been reviewed [31].

Research has also produced modified clotting factors that circulate for longer periods than the native factors; long-acting versions of both recombinant factors VIII and IX have recently been approved by the FDA [32]. These products have the advantage of providing protection against bleeding with fewer intravenous infusions. Although clotting factor replacement with recombinant factors has been the mainstay of hemophilia treatment, an ongoing problem is the development of resistance to the infused proteins. A current investigation is identifying sites on the clotting proteins that elicit the immune responses that destroy the infused therapeutic material [33]. By synthesizing clotting factors that lack these provocative areas, it might be possible to circumvent this treatment resistance. However, this approach will require a great deal of more basic research and probably several clinical trials to confirm the safety and effectiveness of these modified clotting proteins.

The prevention of HIV infection and AIDS is a worldwide problem that requires a global response; research has focused on behavioral modification, male circumcision, and pre-exposure prophylaxis [34]. A major research focus has been the development of a safe and effective vaccine, but this has been difficult because of the extensive variability in the envelope proteins of the virus and the fact that key sites on these proteins are masked from potential antibodies elicited by a vaccine [35]. However, recent research has produced several novel antibodies that are able to neutralize diverse strains of HIV [36]. In addition to assisting in the design of a vaccine, some of these antibodies also target the cells harboring the virus. These cells are the factories for viral synthesis and their elimination might have a long-term beneficial effect on the treatment of the infection. Other investigators are trying to determine why some patients have a natural resistance to HIV infection, and they have discovered that genetic mutations in certain cell membrane proteins thwart the virus from entering those cells. These studies might eventually result in the development of drugs that simulate these mutant proteins and prevent the spread of the infection.

In addition to medical research, there is also a pressing need for behavioral studies focused on ways to change hostile public attitudes toward persons with alternative sexual orientations and those with mental or physical disabilities. In particular, people with hemophilia experience a lack of social support and are under-employed, resulting in a high prevalence of depression [37]. Studies have shown that their quality of life is reduced, but data are lacking on many of the psychosocial aspects of hemophilia [38]. Research is also needed to define the factors that engender hostility and avoidance of those who fall outside of the usual societal norms. The development and early implementation of educational programs focused on instilling tolerance might promote greater acceptance of homosexuals and hemophiliacs. This type of research has been underfunded in the past, but it needs to be addressed if we wish to better the lives of all members of our society.

In summary, this prescription for mitigating the next health care crisis due to blood-borne disease has four components:

- Create an Office of Blood Product Safety (OBPS) within the FDA to alert consumers about threats to the safety of blood and blood products, and ensure that there is an adequate supply of clotting factor concentrates for the needs of persons with bleeding disorders,
- Establish an Office for the Control of Pharmaceutical Prices (COPP) that would place caps on prices for blood, blood products, and essential medications, and authorize subsidies for the manufacture of selected products essential for human health,
- Enhance educational efforts to remove the stigma of infectious disease and instill tolerance toward those that are afflicted, and
- Invest in basic and applied research to improve the care and treatment of those requiring blood and blood products.

KEY POINTS

- The FDA should establish an Office of Blood Product Safety (OBPS) to communicate information on the safety and availability of blood and blood products to the public and health care professionals.
- Blood is freely given and should be freely received; the costs of collection, processing, and distribution should be borne by the public.

- Government should establish an Office for the Control of Pharmaceutical Prices (COPP) to set caps on the prices for blood, blood products, and essential drugs that is modeled after the wartime Office of Price Administration.
- Public investment in education and research will improve the safety and accessibility of blood and blood products for all our citizens.

REFERENCES

[1] WHA28.72 Utilization and supply of human blood and blood products. In: Twenty-eighth World Health Assembly, Geneva; May 13–30, 1975. Available from: <www.who.int/bloodsafety/voluntary_donation/en/>.

[2] Allen JG, Sayman WA. Serum hepatitis from transfusions of blood. JAMA 1962;180:1079–85.

[3] Ness P. The safety of the blood supply. Clin Adv Hematol Oncol 2011;9:607–8.

[4] Leveton LB, Sox Jr. HC, Stoto MA, editors. HIV and the blood supply: an analysis of crisis decisionmaking. Washington, DC: National Academy Press; 1995.

[5] Gallian P, de Lamballerie X, Salez N, et al. Prospective detection of chikungunya virus in blood donors, Caribbean 2014. Blood 2014;123:3679–81.

[6] National Hemophilia Foundation. MASAC recommendation: 225: MASAC recommendations concerning products licensed for the treatment of hemophilia and other bleeding disorders; May 6, 2014.

[7] Wald ML. Blood industry hurt by surplus. NY Times August 23, 2014.

[8] Berry KM. All about blood banks: a multibillion-dollar business in a nonprofit world. NY Times July 7, 1991.

[9] Rosenthal E. Even small medical advances can mean big jumps in bills. NY Times April 6, 2014.

[10] National Institute for Health and Care Excellence. Accessed from: <http://en.wikipedia.org>. Also <http://www.nice.org>; [accessed 09.05.14].

[11] Pollack A. Health insurers pressing down on drug prices. NY Times June 21, 2014.

[12] Thomas K. Cigna agrees to reduce HIV drug costs for some Florida patients. NY Times November 8, 2014.

[13] Alpern JD, Stauffer WM, Kesselheim AS. High-cost generic drugs-implications for patients and policymakers. N Engl J Med 2014;371:1859–62.

[14] Schulman KA, Balu S, Reed SD. Specialty pharmaceuticals for hyperlipidemia-impact on insurance premiums. N Engl J Med 2015;373:1591–3.

[15] Roosevelt FD. Executive Order 8734 establishing the office of price administration and civilian supply. Accessed from: <http://www.presidency.ucsb.edu/ws/?pid=16099>; April 11, 1941.

[16] Servick K. The drug push. Science 2015;348:850–3.

[17] Rosenthal E. For drugs that save lives, a steep cost. NY Times April 27, 2014.

[18] Press release: Cerus announces collaboration agreement with Nipro for innovative intercept red blood cell system component; released April 29, 2014.

[19] FDA Basics, FDA fundamentals. Available from: <www.FDA.gov>; April 28, 2014.

[20] Editorialist. Rise of antibiotic resistance. NY Times May 11, 2014:12.

[21] Quoted by Rick Gladstone in the NY Times November 4, 2014. p. A9.

[22] White R. Testimony before the President's Commission on AIDS; 1988.

[23] Witchel A. At home with Jeanne White-Ginder: a son's AIDS, and a legacy. NY Times September 24, 1992.

[24] Schwarz A, Smith M. Needle exchange is allowed after HIV outbreak in Indiana County. NY Times March 26, 2015.

[25] Associated Press. HIV outbreak in Indiana affecting residents quality of life. NY Times April 21, 2015.

[26] Cooke W. Quoted in Goodnough A. Indiana races to fight HIV surge tied to drug abuse. NY Times March 30, 2015.

[27] Goodnough A. An outbreak fed by misinformation and fear. NY Times May 6, 2015.

[28] Karim SSA. Nelson R Mandela (1918–2013). Science 2014;343:150–1.

[29] Snyder EL, Stramer SL, Benjamin RJ. The safety of the blood supply-time to raise the bar. N Engl J Med 2015;372:1882–4.

[30] Giarratana M-C, Rouard H, Dumont A, et al. Proof of principle for transfusion of in vitro-generated red blood cells. Blood 2011;118:5071–9.

[31] Goodnough LT, Shander A. Current status of pharmacologic therapies in patient blood management. Anesth Analg 2013;116:15–34.

[32] Biogen Idec. Product information for Eloctate and Alprolix.

[33] Nguyen P-CT, Lewis KB, Ettinger RA, et al. High-resolution mapping of epitopes on the C2 domain of factor VIII by analysis of point mutants using surface Plasmon resonance. Blood 2014;123:2732–9.

[34] Piot P, Quinn TC. Response to the AIDS pandemic-a global health model. N Engl J Med 2013;368:2210–18.

[35] Rathore U, Kesavardhana S, Mallajosyula VVA, Varadarajan R. Immunogen design for HIV-1 and influenza. Biochim Biophys Acta 2014;1844:1891–906.

[36] Weiss RA. Immunotherapy for HIV infection. N Engl J Med 2014;370:379–80.

[37] Iannone M, Pennick L, Tom A, et al. Prevalence of depression in adults with haemophilia. Haemophilia 2012;18:868–74.

[38] Cassis FR, Querol F, Forsyth A, Iorio A. HERO International Advisory Board. Psychosocial aspects of haemophilia: a systematic review of methodologies and findings. Haemophilia 2012;18:e101–14.

Further Reading

For readers who wish to delve further into the HIV epidemic, I recommend *And the Band Played On*, by Randy Shilts, a definitive work on the history of the epidemic as it affected the New York and San Francisco homosexual communities in the early 1980s. For more information on hemophilia and AIDS, there are several excellent books. *Blood Saga: Hemophilia, AIDS, and the Survival of a Community* by Susan Resnik consists of in-depth interviews with hemophiliacs, their doctors, and leaders of the National Hemophilia Foundation that are woven together to tell the story of the epidemic. *The Bleeding Disease: Hemophilia and the Unintended Consequences of Medical Progress* by Stephen Pemberton reviews the history of hemophilia treatment and describes some of the adverse effects of technological advances on patient outcomes. A book written from the perspective of the mother of two children infected by HIV is *Cry Bloody Murder: A Tale of Tainted Blood* by Elaine DePrince. The author rails against the inaction by concentrate manufacturers and the Hemophilia Foundation that perpetuated the provision of tainted clotting factor concentrates. *Every Last Drop*, by George T. Baxter, is written from the perspective of an attorney who litigated against the American Association of Blood Banks for not doing enough to protect the blood supply from viral contamination. The responses of governments to the presence of HIV in the blood supply are examined in *Blood Feuds: AIDS, Blood, and the Politics of Medical Disaster*, edited by Eric A Feldman and Ronald Bayer. These several books provide perspectives from many vantage points and provide a detailed history of the epidemic.

Index

Printed in the United States
By Bookmasters